工业和信息化精品系列教材

Java Web
应用开发教程

(项目式)

李文蕙 罗保山 刘嵩 ● 主编
董宁 孙琳 ● 副主编

JAVA WEB APPLICATION
DEVELOPMENT COURSE

人民邮电出版社
北京

图书在版编目（CIP）数据

Java Web 应用开发教程：项目式 / 李文蕙，罗保山，刘嵩主编. -- 北京：人民邮电出版社，2022.7
工业和信息化精品系列教材
ISBN 978-7-115-58692-6

Ⅰ. ①J… Ⅱ. ①李… ②罗… ③刘… Ⅲ. ①JAVA语言-程序设计-高等职业教育-教材 Ⅳ. ①TP312.8

中国版本图书馆CIP数据核字(2022)第027794号

内 容 提 要

本书使用通俗易懂的语言深入浅出地阐述 Java Web 开发前、后端的相关知识，并结合典型的 Web 应用案例，帮助读者掌握 Web 应用程序的开发技术。

本书共 11 章，详细讲解网页开发所需的前端知识和 Java Web 开发的后端知识，其中网页开发的前端知识包括 Bootstrap 框架和 jQuery 框架的使用方法，Java Web 的后端知识包括 JDBC 技术、Servlet 技术、JSP 技术和 JSTL 等。本书在实战篇按照项目开发的流程实现小型超市管理系统，帮助读者更深刻地理解相关知识在项目开发中的使用技巧。

本书适合作为计算机相关专业的教材，也可供广大计算机编程爱好者自学参考。

◆ 主　编　李文蕙　罗保山　刘　嵩
　 副主编　董　宁　孙　琳
　 责任编辑　鹿　征
　 责任印制　王　郁　焦志炜

◆ 人民邮电出版社出版发行　北京市丰台区成寿寺路 11 号
　 邮编　100164　电子邮件　315@ptpress.com.cn
　 网址　https://www.ptpress.com.cn
　 三河市君旺印务有限公司印刷

◆ 开本：787×1092　1/16
　 印张：15.25　　　　　　　2022 年 7 月第 1 版
　 字数：341 千字　　　　　2022 年 7 月河北第 1 次印刷

定价：59.80 元

读者服务热线：(010)81055256　印装质量热线：(010)81055316
反盗版热线：(010)81055315
广告经营许可证：京东市监广登字 20170147 号

前言 FOREWORD

Java 是非常流行的开发语言，在 Web 开发中占据了相当重要的位置。Java Web 开发技术包括 Servlet、JSP、JDBC 及 JSTL 等，是 Java 程序员进行 Web 后端开发的基础技术。近几年随着移动设备的飞速发展，Bootstrap 和 jQuery 也成为前端开发需要掌握的重要技术。

大部分刚开始接触 Java Web 的读者会有这样的问题：感觉什么都学了，又感觉什么都没有学。这是因为大多数人一般会先学 HTML、CSS 和 JavaScript，然后会学习 Java、JSP 和 Servlet，中间还会穿插着数据库的知识，在学习过程中，知识和知识之间缺少联系，学完就忘。市面上虽也不乏一些书籍通过各种大型项目，将知识点全部涵盖。但对于初学者，"大而全"意味着复杂、难以上手。因此编者希望通过一个功能简洁的小型项目将 Java Web 开发需要的核心知识融合在一起，使读者学习起来没有太大压力，让读者按书中步骤一步步做下来可以明白这些知识在项目开发中如何使用。读者掌握知识的基本用法后，自然会通过后续的自主学习去探索更全面的知识体系。

本书共 11 章，第 1~3 章为前端篇，第 4~5 章为后端篇，第 6~11 章为实战篇，具体内容如下。

第 1 章主要讲解 Bootstrap 框架的基础知识，内容包括网格系统、辅助类、表格样式、表单样式和图标的使用。

第 2 章主要讲解 Bootstrap 框架组件和插件的使用，内容包括常用警告框、按钮组、分页、折叠、导航栏和模态框等。

第 3 章主要讲解 jQuery 框架的基础知识，内容包括 jQuery 选择器、jQuery 集合操作、jQuery 事件处理、jQuery 操作 DOM 和 jQuery AJAX。

第 4 章主要讲解 JDBC 的使用，内容包括 JDBC 的基础知识、常用 API、JDBC 事务、PO 和 DAO 的封装思想。

第 5 章主要讲解 Servlet 与 JSP，内容包括 Servlet 运行原理及生命周期、作用域对象、过滤器、JSP 基础、EL 表达式、JSTL、MVC 模式，以及 Servlet 返回 JSON 数据。

第 6 章主要讲解超市管理系统的整体需求、设计，以及基础项目搭建。

第 7~11 章分模块实现超市管理系统的各个功能，内容包括详细需求说明、功能时序图、前端代码和后端代码。

由于编者水平有限，书中难免存在表达欠妥之处，编者由衷希望广大读者朋友和专家学者能够提出宝贵的修改建议。修改建议可直接反馈至编者的电子邮箱：book_javaweb2021@163.com。

编者
2022 年 5 月于武汉

目录 CONTENTS

前端篇

第 1 章 Bootstrap 基础 ············· 1

- 1.1 Bootstrap 概述 ············· 2
  - 1.1.1 什么是 Bootstrap ············· 2
  - 1.1.2 开发工具 ············· 2
  - 1.1.3 下载 Bootstrap ············· 3
  - 1.1.4 第一个 Bootstrap 页面 ············· 4
- 1.2 Bootstrap 网格系统 ············· 5
  - 1.2.1 容器 ············· 5
  - 1.2.2 网格系统 ············· 7
  - 1.2.3 列嵌套 ············· 9
- 1.3 Bootstrap 辅助类 ············· 10
  - 1.3.1 边距 ············· 10
  - 1.3.2 浮动 ············· 12
  - 1.3.3 清除浮动 ············· 13
  - 1.3.4 颜色 ············· 14
  - 1.3.5 边框和阴影 ············· 15
- 1.4 Bootstrap 表格 ············· 16
  - 1.4.1 基础表格 ············· 16
  - 1.4.2 常用表格样式 ············· 18
- 1.5 基础表单 ············· 20
  - 1.5.1 基本用法 ············· 20
  - 1.5.2 常用表单控件 ············· 21
  - 1.5.3 表单网格布局 ············· 23
  - 1.5.4 水平表单 ············· 26
  - 1.5.5 行内表单 ············· 27
- 1.6 图标 ············· 28
  - 1.6.1 Bootstrap 图标下载 ············· 28
  - 1.6.2 Bootstrap 图标使用 ············· 29
- 本章习题 ············· 30

第 2 章 Bootstrap 进阶 ············· 31

- 2.1 Bootstrap 组件 ············· 32
  - 2.1.1 警告框 ············· 32
  - 2.1.2 按钮和按钮组 ············· 34
  - 2.1.3 输入框组 ············· 36
  - 2.1.4 列表组 ············· 38
  - 2.1.5 分页 ············· 39
  - 2.1.6 巨幕 ············· 41
- 2.2 JavaScript 插件 ············· 42
  - 2.2.1 下拉菜单 ············· 42
  - 2.2.2 折叠 ············· 43
  - 2.2.3 导航栏 ············· 45
  - 2.2.4 模态框 ············· 47
- 本章习题 ············· 48

第 3 章 jQuery 基础 ············· 50

- 3.1 jQuery 起步 ············· 51
  - 3.1.1 下载 jQuery ············· 51
  - 3.1.2 第一个 jQuery 程序 ············· 51
- 3.2 jQuery 选择器 ············· 52
  - 3.2.1 jQuery 基本语法 ············· 52
  - 3.2.2 jQuery 选择器 ············· 54
- 3.3 jQuery 集合操作 ············· 55
- 3.4 jQuery 事件处理 ············· 58
- 3.5 jQuery 操作 DOM ············· 61
- 3.6 jQuery AJAX ············· 63
  - 3.6.1 jQuery AJAX 简介 ············· 63
  - 3.6.2 $.get()和$.post() ············· 64
  - 3.6.3 $.getJSON() ············· 66
- 本章习题 ············· 67

后端篇

第 4 章

JDBC ·································· 68

- 4.1 JDBC 基础 ································ 69
  - 4.1.1 什么是 JDBC ···················· 69
  - 4.1.2 环境准备 ························ 69
  - 4.1.3 常用 API ························ 72
  - 4.1.4 JDBC 操作步骤 ·················· 73
  - 4.1.5 第一个 JDBC 程序 ··············· 74
- 4.2 JDBC 基本操作 ·························· 77
  - 4.2.1 PO 和 DAO ······················ 77
  - 4.2.2 JDBC 添加 ······················ 79
  - 4.2.3 JDBC 删除 ······················ 80
  - 4.2.4 JDBC 修改 ······················ 81
  - 4.2.5 JDBC 查询 ······················ 82
  - 4.2.6 JDBC 事务 ······················ 84
- 本章习题 ·· 85

第 5 章

Servlet 与 JSP ······················ 87

- 5.1 Servlet 基础 ······························· 88
  - 5.1.1 环境准备 ························ 88
  - 5.1.2 第一个 Web 应用程序 ·········· 89
  - 5.1.3 Servlet 运行原理 ··············· 92
  - 5.1.4 Servlet 生命周期 ··············· 93
  - 5.1.5 Servlet 请求和响应 ············ 94
  - 5.1.6 乱码处理 ························ 98
  - 5.1.7 重定向与转发 ···················· 99
  - 5.1.8 作用域与数据共享 ············· 101
- 5.2 JSP 基础 ································· 103
  - 5.2.1 JSP 运行原理 ················· 103
  - 5.2.2 JSP 内置对象 ················· 103
  - 5.2.3 JSP 标签 ······················· 104
- 5.3 EL 表达式和 JSTL ······················ 106
  - 5.3.1 EL 表达式 ····················· 106
  - 5.3.2 JSTL core 标签库 ············ 108
- 5.4 过滤器 ···································· 112
  - 5.4.1 过滤器简介 ····················· 112
  - 5.4.2 过滤器使用 ····················· 113
- 5.5 MVC 模式 ······························· 113
- 5.6 Servlet 返回 JSON 数据 ·············· 118
- 本章习题 ······································· 120

实战篇

第 6 章

项目需求与设计 ······················ 121

- 6.1 项目概述 ································· 122
- 6.2 系统设计 ································· 123
  - 6.2.1 实体类设计 ····················· 123
  - 6.2.2 数据库设计 ····················· 124
  - 6.2.3 页面原型设计 ·················· 125
- 6.3 项目准备 ································· 126
  - 6.3.1 项目搭建 ························ 126
  - 6.3.2 基础页面实现 ·················· 127
  - 6.3.3 实体类实现 ····················· 129
  - 6.3.4 工具类实现 ····················· 130
  - 6.3.5 登录和退出实现 ··············· 131
  - 6.3.6 权限功能实现 ·················· 135
- 本章习题 ······································· 136

第 7 章

供应商管理模块实现 ················ 137

- 7.1 查询供应商 ······························ 138
  - 7.1.1 任务需求 ························ 138
  - 7.1.2 任务实现 ························ 140
- 7.2 添加供应商 ······························ 143
  - 7.2.1 任务需求 ························ 143
  - 7.2.2 任务实现 ························ 145
- 7.3 删除供应商 ······························ 149
  - 7.3.1 任务需求 ························ 149
  - 7.3.2 任务实现 ························ 150
- 7.4 修改供应商 ······························ 152
  - 7.4.1 任务需求 ························ 152

7.4.2　任务实现 …………………… 154
本章习题 …………………………………… 159

第 8 章

分类管理模块实现 …………………… 160

8.1　查询分类 …………………………… 161
　　8.1.1　任务需求 …………………… 161
　　8.1.2　任务实现 …………………… 163
8.2　添加分类 …………………………… 166
　　8.2.1　任务需求 …………………… 166
　　8.2.2　任务实现 …………………… 168
8.3　删除分类 …………………………… 171
　　8.3.1　任务需求 …………………… 171
　　8.3.2　任务实现 …………………… 172
8.4　修改分类 …………………………… 174
　　8.4.1　任务需求 …………………… 174
　　8.4.2　任务实现 …………………… 176
本章习题 …………………………………… 180

第 9 章

商品管理模块实现 …………………… 181

9.1　查询商品 …………………………… 182
　　9.1.1　任务需求 …………………… 182
　　9.1.2　任务实现 …………………… 183
9.2　添加商品 …………………………… 186
　　9.2.1　任务需求 …………………… 186
　　9.2.2　任务实现 …………………… 188
9.3　删除商品 …………………………… 193
　　9.3.1　任务需求 …………………… 193

　　9.3.2　任务实现 …………………… 194
9.4　修改商品 …………………………… 196
　　9.4.1　任务需求 …………………… 196
　　9.4.2　任务实现 …………………… 198
本章习题 …………………………………… 204

第 10 章

进货管理模块实现 …………………… 205

10.1　添加进货记录 ……………………… 206
　　10.1.1　任务需求 ………………… 206
　　10.1.2　任务实现 ………………… 208
10.2　查询进货记录 ……………………… 213
　　10.2.1　任务需求 ………………… 213
　　10.2.2　任务实现 ………………… 215
本章习题 …………………………………… 218

第 11 章

销售模块实现 ………………………… 219

11.1　添加销售记录 ……………………… 220
　　11.1.1　任务需求 ………………… 220
　　11.1.2　任务实现 ………………… 222
11.2　查询销售记录 ……………………… 228
　　11.2.1　任务需求 ………………… 228
　　11.2.2　任务实现 ………………… 229
11.3　查看销售明细 ……………………… 232
　　11.3.1　任务需求 ………………… 232
　　11.3.2　任务实现 ………………… 234
本章习题 …………………………………… 236

前 端 篇

第1章
Bootstrap基础

01

本章目标

- 理解 Bootstrap 框架
- 掌握 Bootstrap 网格系统
- 掌握 Bootstrap 辅助类
- 掌握 Bootstrap 表格样式
- 掌握 Bootstrap 表单样式
- 掌握 Bootstrap 图标

1.1 Bootstrap 概述

1.1.1 什么是 Bootstrap

Bootstrap 是一种基于 HTML、CSS 和 JavaScript 的前端开发框架，它常被用来开发响应式、移动设备优先的网站。Bootstrap 具有以下优点。

（1）跨浏览器。Bootstrap 能够兼容目前的主流浏览器，开发人员在使用 Bootstrap 时不用再考虑如何解决 IE 浏览器和其他浏览器之间的兼容性问题，降低了开发难度。

（2）响应式布局。Bootstrap 的 CSS 类是基于响应式布局构建的，这让同一套代码可以根据浏览设备的分辨率变化调整显示内容。

（3）自带丰富的组件。Bootstrap 自带了常用的页面组件，例如导航栏、表单控件、分页等，提高了开发效率。

（4）提供 JavaScript 插件。Bootstrap 提供了实用性很强的 JavaScript 插件，使用这些插件，开发人员可以快速实现多种特效，例如下拉菜单、模态框、轮播等。

（5）支持 CSS 预编译。Bootstrap 支持 Less、Sass 等预编译语言，通过变量、嵌套、混合、函数等功能可以开发出灵活多变的页面。Bootstrap 中很多样式的属性值都是通过预编译语言计算得到的。通过浏览器调试工具直接查看元素得到的是计算结果，如果想看原始代码，需要查看对应的.scss 文件。

1.1.2 开发工具

HBuilder 是一款优秀的国产开发工具，具有免费、轻巧、极速、跨平台等特点。用户可以在 HBuilder 官网下载 HBuilder X，如图 1-1 所示。下载时可以选择标准版，HBuilder 标准版可直接用于 Web 开发，相对于 App 开发版更小巧。

图 1-1　HBuilder 下载界面

1.1.3 下载 Bootstrap

Bootstrap 目前最新的版本是 2021 年 8 月发布的 5.1.0 版本，距离 5.0 版本发布仅过了 3 个月。本书使用更稳定的 Bootstrap4.6 版本。

可以通过以下 3 种方式将 Bootstrap 引入项目中使用。

（1）下载已经编译好的代码到本地，然后在项目中直接引用。

（2）通过 jsDelivr 服务器直接引用网络 CSS 和 JS 文件到项目中。

（3）通过各种包管理器进行下载。

第一种方式简单直观，下载后可以脱离网络使用 Bootstrap，第二种方式适合项目发布后进行网络镜像加速，第三种方式适合工程化项目开发。本书以第一种方式为例进行介绍。

访问 Bootstrap 官网，打开页面单击"All releases"链接，跳转到版本选择，如图 1-2 所示。单击 4.6 版本的链接，打开 Bootstrap4.6 版本主页面，如图 1-3 所示；单击"Head to the downloads page."链接跳转到下载页面，单击"Compiled CSS and JS"处的"Download"按钮下载，如图 1-4 所示。

图 1-2　Bootstrap 官网

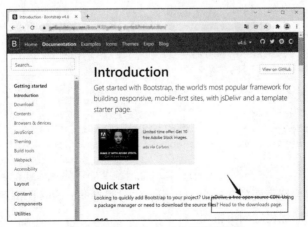

图 1-3　Bootstrap 4.6 版本主页面

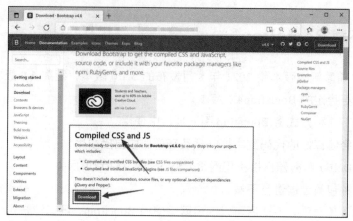

图 1-4 下载 Bootstrap

下载的压缩包名为"bootstrap-4.6.0-dist.zip",解压后得到 css 和 js 两个文件夹。其中扩展名为.min 的文件是精简文件,删除了多余的空格和换行符,压缩了代码体积。

1.1.4　第一个 Bootstrap 页面

在使用 Bootstrap 时必须在页面中引入 Bootstrap.css,如果想使用 Bootstrap 提供的插件,还需要引入 Bootstrap.js。由于 Bootstrap4.6 中的插件都依赖 jQuery,所以在引入 Bootstrap.js 前还需要引入 jQuery.js。可以通过 jQuery 官网下载 jquery-3.6.0.min.js。

打开 HBuilder X 后单击"新建项目"按钮,在"新建"对话框中输入项目名"chap01",选择模板"基本 HTML 项目"。项目新建完成后将 Bootstrap 压缩包解压出来的 css 和 js 文件夹中的内容复制到项目对应的目录下,并将下载的 jquery-3.6.0.min.js 也复制到项目的 js 文件夹下。

【案例 1-1】index.html

```
<!DOCTYPE html>
<html lang="en">
  <head>
    <meta charset="utf-8">
    <meta name="viewport" content="width=device-width, initial-scale=1, shrink-to-fit=no">
    <link rel="stylesheet" href="css/bootstrap.css">
    <title>Hello, world!</title>
  </head>
  <body>
    <h1>Hello, world!</h1>
    <script src="js/jquery-3.6.0.min.js"></script>
    <script src="js/bootstrap.bundle.js"></script>
  </body>
</html>
```

移动设备浏览器会把页面放在一个 viewport 中,通常 viewport 比屏幕宽,这样会破坏没有针对移动设备浏览器优化的网页布局。

代码<meta name="viewport" content="width=device-width, initial-scale=1, shrink-to-fit=no">用于设置页面显示的宽度为设备宽度，初始不缩放，可以让网页的宽度自动适应移动设备屏幕的宽度。

代码<link rel="stylesheet" href="css/bootstrap.css">用于引入 Bootstrap 的样式。

代码末尾先后引入了 jquery-3.6.0.min.js 和 bootstrap.bundle.js，如果不需要使用特效，这两个文件可以不引入；但只要引入，它们的先后顺序是不能颠倒的，否则会报错。

代码编辑完毕后单击"运行"按钮，并选择合适的浏览器，如图 1-5 所示。运行成功会得到结果，如图 1-6 所示。

图 1-5　选择浏览器

图 1-6　运行结果

运行结果看起来与原生 HTML 没什么区别，在浏览器页面按【F12】键打开调试模式，在元素区域中选中 h1，在右侧的 Style 区域可以看见这个元素附加了多个样式，修改了字体大小、间距等内容。

1.2　Bootstrap 网格系统

1.2.1　容器

容器（.container）是 Bootstrap 中最基本的布局样式，Bootstrap 提供了以下 3 种容器样式。

（1）.container：带有最大宽度的响应式容器。它会跟随设备宽度变化调整最大宽度，大多数情况会在页面两侧留下空白区域。

（2）.container-{设备}：如果设备宽度小于指定宽度，则容器的最大宽度为 100%，超过指定宽度则最大宽度为指定宽度。设备的值有 sm（屏幕宽度为 576～768px）、md（屏幕宽度为 768～992px）、lg（屏幕宽度为 992～1200px）和 xl（屏幕宽度在 1200px 以上）。

（3）.container-fluid：最大宽度为 100%的容器。

Bootstrap 根据设备的屏幕宽度不同将其分为 5 个级别，不同容器样式下 viewport 的最大宽度和设备宽度的对应关系如表 1-1 所示。

表 1-1　容器、设备宽度对应表

	极小型 <576px	小型 ≥576px且 <768px	中型 ≥768px且 <992px	大型 ≥992px且 ≤1200px	极大型 >1200px
.container	100%	540px	720px	960px	1140px
.container-sm	100%	540px	720px	960px	1140px
.container-md	100%	100%	720px	960px	1140px
.container-lg	100%	100%	100%	960px	1140px
.container-xl	100%	100%	100%	100%	1140px
.container-fluid	100%	100%	100%	100%	100%

【案例 1-2】 1-1.html

```html
<!DOCTYPE html>
<html lang="en">
  <head>
    <meta charset="utf-8">
    <meta name="viewport" content="width=device-width, initial-scale=1, shrink-to-fit=no">
    <link rel="stylesheet" href="css/bootstrap.css">
    <style>
    div{
      border: 1px solid;
    }
    </style>
  </head>
  <body>
    <div class="container">
      根据设备屏幕宽度调整最大宽度的容器
    </div>
    <div class="container-fluid">
      100%设备屏幕宽度的容器
    </div>
    <div class="container-sm">100% 宽度直到小型设备屏幕</div>
    <div class="container-md">100% 宽度直到中型设备屏幕</div>
    <div class="container-lg">100% 宽度直到大型设备屏幕</div>
    <div class="container-xl">100% 宽度直到极大型设备屏幕</div>
  </body>
</html>
```

代码运行结果如图 1-7 所示，调整浏览器的宽度可以看到容器宽度在发生变化。

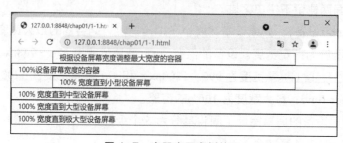

图 1-7　容器应用案例效果

1.2.2 网格系统

Bootstrap 提供了一套响应式、移动设备优先的网格系统。这套系统根据屏幕宽度变化自动划分为最多 12 列。

响应式网格系统使用容器（.container）、行（.row）和单元格（.col-{xx}-{xx}）布局。

网格系统提供了 5 个类来进行单元格响应，分别是.col-（极小设备）、.col-sm-（小型设备）、.col-md-（中型设备）、.col-lg-（大型设备）和.col-xl-（极大型设备）。

【案例 1-3】1-2.html

```html
<!DOCTYPE html>
<html lang="en">
  <head>
    <meta charset="utf-8">
    <meta name="viewport" content="width=device-width, initial-scale=1, shrink-to-fit=no">
    <link rel="stylesheet" href="css/bootstrap.css">
    <style>
    div{
      border: 1px solid;
    }
    </style>
  </head>
  <body>
    <div class="container">
      <div class="row">
        <div class="col-md-2">col-md-2</div>
        <div class="col-md-4">col-md-4</div>
        <div class="col-md-6">col-md-6</div>
      </div>
      <div class="row">
        <div class="col-lg-2">col-lg-2</div>
        <div class="col-lg-4">col-lg-4</div>
        <div class="col-lg-6">col-lg-6</div>
      </div>
    </div>
  </body>
</html>
```

网格系统的基本代码结构如案例 1-3 所示，一个容器中可以包含多个行，每一个行可以包含多个单元格，每一个行会根据其直接子元素单元格划分为最多 12 列。单元格中可以指定设备宽度，如果设备宽度小于指定宽度，则每个单元格独占 12 列；如果设备宽度大于指定宽度，则按指定的数字分配宽度。

上面代码的运行结果如图 1-8 所示。调整屏幕宽度为 600px（小型设备宽度 sm）、800px（中型设备宽度 md）和 1200px（极大型设备宽度 xl）。当设备宽度为 sm 级别时，.col-md 和.col-lg 都是独占一行。当设备宽度为 md 级别时，.col-md 会根据其后的数字分配宽度，而.col-lg 还是独占一行。当设备宽度为 xl 级别时，.col-md 和.col-lg 都按照后面的数字分配宽度。

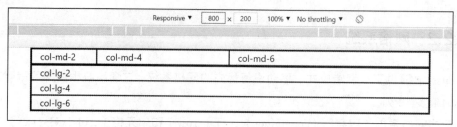

图1-8 屏幕宽度为800px时的运行结果

有时希望简化单元格布局的代码,可以使用下面的简写方式。

(1)如果希望单元格在所有屏幕宽度上的布局保持一致,可以直接使用.col-{数字}来指定宽度,数字值为1~12,这样同一个单元格占的网格列数在不同分辨率下会保持一致。

(2)如果希望单元格平均分配宽度,可以直接使用.col,这样网格系统会自动根据一行中的.col个数计算宽度。

(3)如果希望一行中部分单元格指定占用列数,剩下的单元格平均分配,可以组合使用.col和.col-{数字}。

【案例1-4】1-3.html

```html
<!DOCTYPE html>
<html lang="en">
  <head>
    <meta charset="utf-8">
    <meta name="viewport" content="width=device-width, initial-scale=1, shrink-to-fit=no">
    <link rel="stylesheet" href="css/bootstrap.css">
    <style>
    div{
      border: 1px solid;
    }
    </style>
  </head>
  <body>
    <div class="container">
      <div class="row">
        <div class="col-2">col-2</div>
        <div class="col-4">col-4</div>
        <div class="col-6">col-6</div>
      </div>
      <div class="row">
        <div class="col">col</div>
        <div class="col">col</div>
        <div class="col">col</div>
      </div>
      <div class="row">
        <div class="col">col</div>
        <div class="col">col</div>
        <div class="col">col</div>
        <div class="col">col</div>
        <div class="col">col</div>
      </div>
```

```html
        <div class="row">
          <div class="col">col</div>
          <div class="col-6">col-6</div>
          <div class="col">col</div>
        </div>
      </div>
   </body>
</html>
```

代码运行结果如图 1-9 所示，其中没有指定宽度的单元格会被平均分配剩下的宽度，例如第 2 行是三等分，第 3 行是五等分；第 4 行中间单元格被指定为 6 个列宽，则剩下左右两边单元格各占 3 个列宽。

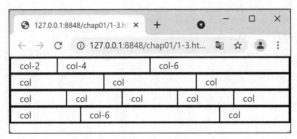

图 1-9 简化网格

1.2.3 列嵌套

Bootstrap 允许在单元格中再次嵌套网格，可以添加一个新的 .row 元素和一系列 .col 元素到已经存在的单元格元素内。

【案例 1-5】1-4.html

```html
<!DOCTYPE html>
<html lang="en">
  <head>
    <meta charset="utf-8">
    <meta name="viewport" content="width=device-width, initial-scale=1, shrink-to-fit=no">
    <link rel="stylesheet" href="css/bootstrap.css">
    <style>
    div{
      border: 1px solid;
    }
    </style>
  </head>
  <body>
    <div class="container">
      <div class="row">
        <div class="col-8">
          col-8
          <div class="row">
            <div class="col-8 col-md-6">
              嵌套层：.col-8 .col-md-6
            </div>
```

```html
            <div class="col-4 col-md-6">
                嵌套层：.col-4 .col-md-6
            </div>
          </div>
        </div>
      </div>
      <div class="row">
        <div class="col-lg-9">
          .col-lg-9
          <div class="row">
            <div class="col col-md-4">
                嵌套层：.col .col-md-4
            </div>
            <div class="col col-md-8">
                嵌套层：.col .col-md-8
            </div>
          </div>
        </div>
      </div>
    </div>
  </body>
</html>
```

代码运行结果如图 1-10 所示，可以调整屏幕宽度看看结果有什么变化。

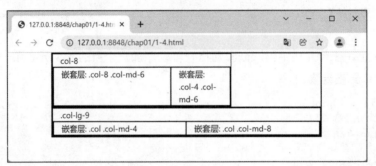

图 1-10　网格嵌套

1.3　Bootstrap 辅助类

1.3.1　边距

Bootstrap 提供了一系列的 CSS 类来调整元素的外边距 margin 和内边距 padding。其语法为 .{类型}{方向}-{大小} 或者 .{类型}{方向}-{设备}-{大小}，具体含义如下。

类型可用值：m 和 p。m 表示设置 margin 的值，p 表示设置 padding 的值。

方向可用值：t、b、l、r、x 和 y。t 表示 top（顶部），b 表示 bottom（底部），l 表示 left（左边），r 表示 right（右边），x 表示设置左右两边，y 表示设置上下两边。

设备可用值：sm、md、lg 和 xl。含义同网格系统，屏幕宽度超过设定值时边距设置起效果。

大小可用值：0～5 和 auto。如果指定数字，会根据 Sass 中 $spacer 的值进行计算。

【案例 1-6】1-5.html

```html
<!DOCTYPE html>
<html lang="en">
  <head>
    <meta charset="utf-8">
    <meta name="viewport" content="width=device-width, initial-scale=1, shrink-to-fit=no">
    <link rel="stylesheet" href="css/bootstrap.css">
    <style>
      div {
        border: 1px solid;
      }
    </style>
  </head>
  <body>
    <div>
      <div class="p-0 mt-4">
        p-0 mt-4
      </div>
      <div class="py-5 mt-0">
        py-5 mt-0
      </div>
      <div class="px-2 my-3">
        px-2 my-3
      </div>
      <div class="pl-4 mx-5">
        pl-4 mx-5
      </div>
    </div>
  </body>
</html>
```

代码运行结果如图 1-11 所示。

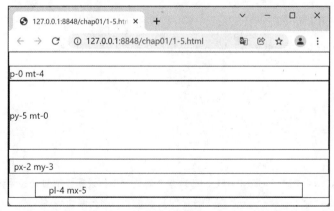

图 1-11 边距

利用 .mx-auto 样式可以让具有固定 Width（宽度）的块元素水平居中。

【案例 1-7】 1-6.html

```html
<!DOCTYPE html>
<html lang="en">
  <head>
    <meta charset="utf-8">
    <meta name="viewport" content="width=device-width, initial-scale=1, shrink-to-fit=no">
    <link rel="stylesheet" href="css/bootstrap.css">
    <style>
      div {
        background-color: lightgray;
      }
    </style>
  </head>
  <body>
    <div class="mx-auto" style="width: 400px;">
      具有固定 width 的块元素可以通过.mx-auto 实现水平居中
    </div>
  </body>
</html>
```

1.3.2 浮动

Bootstrap 提供了一系列的 CSS 类来控制元素浮动。其样式语法为.float-{方式}或者.float-{设备}-{方式}，具体含义如下。

设备可用值：sm、md、lg 和 xl。含义同网格系统，屏幕宽度超过设定值时边距设置起效果。

方式可用值：left、right 和 none。left 表示向左浮动，right 表示向右浮动，none 表示取消浮动。

【案例 1-8】 1-7.html

```html
<!DOCTYPE html>
<html lang="en">
  <head>
    <meta charset="utf-8">
    <meta name="viewport" content="width=device-width, initial-scale=1, shrink-to-fit=no">
    <link rel="stylesheet" href="css/bootstrap.css">
    <style>
      div {
        border: 1px solid;
      }
    </style>
  </head>
  <body>
    <div class="float-md-left">在屏幕宽度达到中型设备以上时向左浮动</div><br />
    <div class="float-right">在任意尺寸设备上都向右浮动</div><br />
    <div class="float-none">不浮动的元素</div>
  </body>
</html>
```

代码运行结果如图 1-12 所示，调整浏览器宽度，发现第一个层只有在宽度超过 md 时才向左浮动，如果浏览器宽度不够则不浮动。

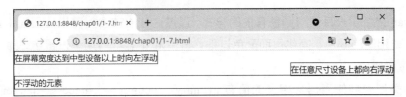

图 1-12　浮动

1.3.3　清除浮动

如果一个元素内部的子元素都是浮动的，会发生父元素不能被子元素撑开的情况，案例如下。

【案例 1-9】1-8.html

```html
<!DOCTYPE html>
<html lang="en">
  <head>
    <meta charset="utf-8">
    <meta name="viewport" content="width=device-width, initial-scale=1, shrink-to-fit=no">
    <link rel="stylesheet" href="css/bootstrap.css">
    <style>
      div {
        border: 1px solid;
      }
    </style>

  </head>
  <body>
    <div style="background-color: lightgray;" class="">
      <div class="float-left">左浮动</div>
      <div class="float-right">右浮动</div>
    </div>
  </body>
</html>
```

Bootstrap 提供了 .clearfix 样式来解决这个问题。为上面代码的父元素加入 class="clearfix"，运行结果如图 1-13 所示，父元素内容正常显示。

图 1-13　清除浮动

1.3.4 颜色

Bootstrap 提供了一系列与颜色有关的样式，通过这些 CSS 类，可以让页面中具有相似概念的元素呈现相同的颜色，简化页面配色过程。其中.text-{类型}用来设置字体颜色，.bg-{类型}用来设置背景色。

常用的类型中与使用语义相关的值有 primary、secondary、success、danger、warning、info、dark、light 和 white。另外字体颜色还有.text-muted、.text-white-50 和.text-black-50 样式。

【案例 1-10】1-9.html

```html
<!DOCTYPE html>
<html lang="en">
  <head>
    <meta charset="utf-8">
    <meta name="viewport" content="width=device-width, initial-scale=1, shrink-to-fit=no">
    <link rel="stylesheet" href="css/bootstrap.css">
    <style>
      div{
        width: 33%;
        display: inline-block;
      }
    </style>
  </head>
  <body>
    <div class="p-3 mb-2 bg-primary text-white">bg-primary text-white</div>
    <div class="p-3 mb-2 bg-secondary text-white">bg-secondary text-white</div>
    <div class="p-3 mb-2 bg-success text-white">bg-success text-white</div>
    <div class="p-3 mb-2 bg-danger text-white">bg-danger text-white</div>
    <div class="p-3 mb-2 bg-warning text-white">bg-warning text-white</div>
    <div class="p-3 mb-2 bg-info text-white">bg-info text-white</div>
    <div class="p-3 mb-2 bg-info text-white">bg-info text-white</div>
    <div class="p-3 mb-2 bg-dark text-light">bg-dark text-light</div>
    <div class="p-3 mb-2 bg-white text-primary">bg-white text-primary</div>
    <div class="p-3 mb-2 bg-white text-secondary">bg-white text-secondary</div>
    <div class="p-3 mb-2 bg-white text-success">bg-white text-success</div>
    <div class="p-3 mb-2 bg-white text-danger">bg-white text-danger</div>
    <div class="p-3 mb-2 bg-white text-warning">bg-white text-warning</div>
    <div class="p-3 mb-2 bg-white text-info">bg-white text-info</div>
    <div class="p-3 mb-2 bg-transparent text-black-50">bg-transparent text-black-50</div>
  </body>
</html>
```

代码运行结果如图 1-14 所示。

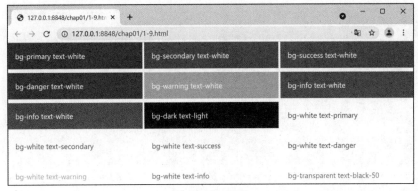

图 1-14　颜色

1.3.5　边框和阴影

Bootstrap 提供了一系列跟边框有关的样式。这些样式可以快速为元素添加边框，设置边框颜色、边框圆角和阴影效果等。

.border：设置整体边框。

.border-{方向}：设置指定方向边框。

border-{方向}-0：取消指定方向边框。方向的可用值有 top、right、bottom、left，top 表示顶部边框，right 表示右侧边框，bottom 表示底部边框，left 表示左侧边框。

.border-{颜色}：设置边框颜色。颜色的可用值有 primary、secondary、success、danger、warning、info、light、dark 和 white。

.border-0：取消边框。

.rounded：设置整体圆角。

.rounded-{方向}：设置指定方向的圆角。方向的可用值有 top、right、bottom、left，top 表示顶部两个角，right 表示右侧两个角，bottom 表示底部两个角，left 表示左侧两个角。

.rounded-{效果}：设置圆角样式。效果可用值有 pill、circle，pill 表示胶囊效果，circle 表示圆形效果。

.rounded-{大小}：设置圆角率。大小的可用值有 lg 和 sm，lg 表示大圆角，sm 表示小圆角。

.rounded-0：取消圆角。

.shadow：设置标准阴影。

.shadow-{大小}：设置指定大小的阴影。大小可用值有 lg 和 sm，lg 表示大号阴影，sm 表示小号阴影。

.shadow-none：取消阴影。

【案例 1-11】1-10.html

```
<!DOCTYPE html>
<html lang="en">
  <head>
    <meta charset="utf-8">
```

```
    <meta name="viewport" content="width=device-width, initial-scale=1, shrink-to-fit=no">
    <link rel="stylesheet" href="css/bootstrap.css">
    <style>
      span{
        display: inline-block;
      }
    </style>
  </head>
  <body>
    <span class="p-3 m-2 border border-primary">border border-primary</span>
    <span class="p-3 m-2 border border-bottom-0 border-danger">border border-bottom-0 border-danger</span>
    <span class="p-3 m-2 border-top border-bottom border-info">border-top border-bottom border-info</span><br>
    <span class="p-3 m-2 border border-success rounded">border border-success rounded</span>
    <span class="p-3 m-2 border border-success rounded-right">border border-success rounded-right</span>
    <span class="p-3 m-2 border border-success rounded-lg">border border-success rounded-lg</span>
    <span class="p-3 m-2 shadow">shadow</span>
  </body>
</html>
```

代码运行结果如图 1-15 所示。

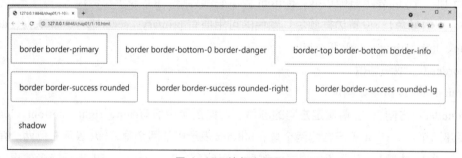

图 1-15 边框和阴影

1.4 Bootstrap 表格

日常开发中经常使用表格来展示多条数据，Bootstrap 中加入了一系列表格美化样式。

1.4.1 基础表格

.table 用于设置表格的基础样式，其他表格样式需要搭配 .table 样式一起使用。

【案例 1-12】1-11.html

```
<!DOCTYPE html>
<html lang="en">
  <head>
```

```html
    <meta charset="utf-8">
    <meta name="viewport" content="width=device-width, initial-scale=1, shrink-to-fit=no">
    <link rel="stylesheet" href="css/bootstrap.css">
  </head>
  <body>
    <table class="table">
      <thead>
        <tr>
          <th scope="col">编号</th>
          <th scope="col">供应商</th>
          <th scope="col">联系人</th>
          <th scope="col">联系电话</th>
        </tr>
      </thead>
      <tbody>
        <tr>
          <th scope="row">1</th>
          <td>供应商 1</td>
          <td>联系人 1</td>
          <td>13980000001</td>
        </tr>
        <tr>
          <th scope="row">2</th>
          <td>供应商 2</td>
          <td>联系人 2</td>
          <td>13980000002</td>
        </tr>
        <tr>
          <th scope="row">3</th>
          <td>供应商 3</td>
          <td>联系人 3</td>
          <td>13980000003</td>
        </tr>
        <tr>
          <th scope="row">4</th>
          <td>供应商 4</td>
          <td>联系人 4</td>
          <td>13980000004</td>
        </tr>
      </tbody>
    </table>
  </body>
</html>
```

代码运行结果如图 1-16 所示，通过调试模式可以看到 .table 样式给原始表格设置了边距、部分边框和字体颜色等。

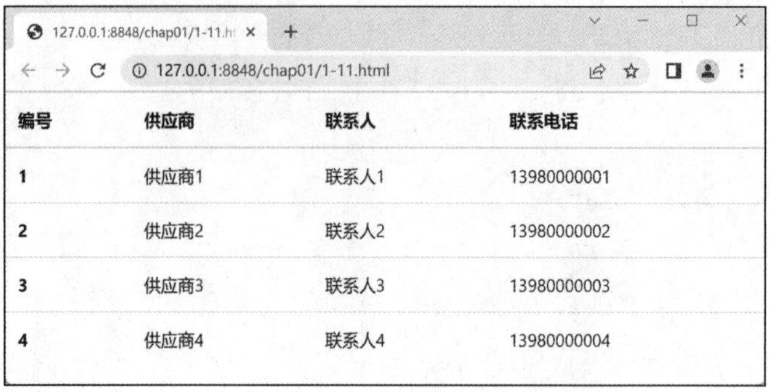

图 1-16 基础表格

1.4.2 常用表格样式

除了基础表格样式，Bootstrap 还提供了表格美化样式。下面的样式在<table>标签上使用，对整个表格起作用。

.table-{颜色}：设置表格为深色背景、浅色字体。

.table-striped：生成逐行颜色对比的表格样式。

.table-bordered：生成表格边框。

.table-borderless：取消表格边框。

.table-hover：生成鼠标指针悬停效果。

.table-sm：生成更紧凑的表格。

还有些样式是在<tr>、<th>或者<td>标签上使用，对其所在的元素起作用，用来设置背景色，包括.table-active、.table-primary、.table-secondary、.table-success、.table-danger、.table-warning、.table-info、.table-light 和.table-dark。

.thead-light 或.thead-dark 在<thead>标签上使用，使<thead>背景显示为浅黑色或深灰色。

【案例 1-13】1-12.html

```
<!DOCTYPE html>
<html lang="en">
  <head>
    <meta charset="utf-8">
    <meta name="viewport" content="width=device-width, initial-scale=1, shrink-to-fit=no">
    <link rel="stylesheet" href="css/bootstrap.css">
  </head>
  <body>
    <table class="table table-sm table-hover table-striped table-bordered">
      <thead class="thead-dark">
        <tr>
          <th scope="col">编号</th>
          <th scope="col">供应商</th>
          <th scope="col">联系人</th>
          <th scope="col">联系电话</th>
```

```html
            </tr>
        </thead>
        <tbody>
            <tr>
                <th scope="row">1</th>
                <td>供应商 1</td>
                <td>联系人 1</td>
                <td>13980000001</td>
            </tr>
            <tr>
                <th scope="row">2</th>
                <td>供应商 2</td>
                <td>联系人 2</td>
                <td>13980000002</td>
            </tr>
            <tr>
                <th scope="row">3</th>
                <td>供应商 3</td>
                <td>联系人 3</td>
                <td>13980000003</td>
            </tr>
            <tr class="table-danger">
                <th scope="row">4</th>
                <td>供应商 4</td>
                <td>联系人 4</td>
                <td>13980000004</td>
            </tr>
            <tr>
                <th scope="row" class="table-info">5</th>
                <td>供应商 5</td>
                <td class="table-success">联系人 5</td>
                <td>13980000005</td>
            </tr>
        </tbody>
    </table>
  </body>
</html>
```

代码运行结果如图 1-17 所示。

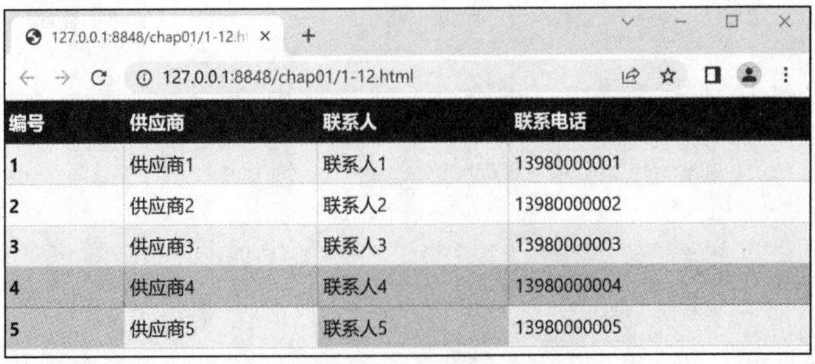

图 1-17 应用多种样式的表格

1.5 基础表单

表单是 Web 开发中必不可少的元素，Bootstrap 提供了一系列 CSS 类来帮助表单进行排版及对表单控件进行美化。

1.5.1 基本用法

Bootstrap 表单是通过在原生表单标签上应用各种 CSS 类来实现的。这些 CSS 类可以使表单在不同的浏览器上呈现一致的效果。

表单的基本用法如下。

```html
<form>
    <div class="form-group">
      <label>控件描述</label>
        <input type="控件类型" class="form-control" >
      </div>
</form>
```

在<form>标签内部添加 class="form-group"的<div>标签，.form-group 为这个 div 的底部添加 1rem 的外边距。在这个 div 中依次添加<label>标签来描述控件，通过 class="form-control"的<input>标签来添加一个控件。.form-control 会对控件的背景、边框、阴影、大小等内容进行设置。

【案例 1-14】1-13.html

```html
<!DOCTYPE html>
<html lang="en">
  <head>
    <meta charset="utf-8">
    <meta name="viewport" content="width=device-width, initial-scale=1, shrink-to-fit=no">
    <link rel="stylesheet" href="css/bootstrap.css">
  </head>
  <body>
    <div class="mx-auto my-5 p-5 shadow" style="width: 600px;">
      <form>
        <div class="form-group">
          <label for="account">账号</label>
          <input type="text" class="form-control" id="account" >
        </div>
        <div class="form-group">
          <label for="pwd">密码</label>
          <input type="password" class="form-control" id="pwd">
        </div>
        <button type="submit" class="btn btn-primary">登录</button>
      </form>
    </div>
  </body>
</html>
```

代码运行结果如图 1-18 所示，原生表单加上几个简单样式就被美化了不少。

1.5.2 常用表单控件

常用的表单控件标签，如<input>、<textarea>、<select>，可以直接使用.form-control 样式，但有一些建议使用特定的 CSS 类以修复一些显示问题，例如文件选择控件<input type="file">使用.form-control-file 样式，范围选择控件<input type="range">使用.form-control-range 样式。

图 1-18 基础表单

单选按钮和复选框代码结构和前面的有所区别，如下所示。

```
<form>
    <div class="form-check">
        <input type="radio" class="form-check-input" >
      <label class="form-check-label" >单选按钮</label>
    </div>
    <div class="form-check">
        <input type="checkbox" class="form-check-input" >
      <label class="form-check-label" >复选框</label>
    </div>
</form>
```

单选按钮或复选框的父元素<div>使用.form-check 样式，单选按钮或者复选框使用.form-check-input 样式，描述<label>需要加上.form-check-label 样式。

【案例 1-15】1-14.html

```
<!DOCTYPE html>
<html lang="en">
  <head>
    <meta charset="utf-8">
    <meta name="viewport" content="width=device-width, initial-scale=1, shrink-to-fit=no">
    <link rel="stylesheet" href="css/bootstrap.css">
  </head>
  <body>
    <div class="mx-auto my-5 p-5 shadow" style="width: 600px;">
      <form>
        <div class="form-group">
          <label>文本框</label>
          <input type="text" class="form-control">
        </div>
        <div class="form-group">
          <label>密码框</label>
          <input type="password" class="form-control">
        </div>
        <div class="form-group">
          <label>下拉列表</label>
          <select class="form-control">
```

```html
        </select>
      </div>
      <div class="form-group">
        <label>文本区</label>
        <textarea class="form-control"></textarea>
      </div>
      <div class="form-group">
        <label>范围选择</label>
        <input type="range" class="form-control-range">
      </div>
      <div class="form-group">
        <label>文件选择</label>
        <input type="file" class="form-control-file">
      </div>
      <div class="form-check">
        <input type="radio" class="form-check-input" />
        <label class="form-check-label">单选按钮</label>
      </div>
      <div class="form-check">
        <input type="checkbox" class="form-check-input" />
        <label class="form-check-label">复选框</label>
      </div>
    </form>
  </div>
</body>
</html>
```

代码运行结果如图 1-19 所示。

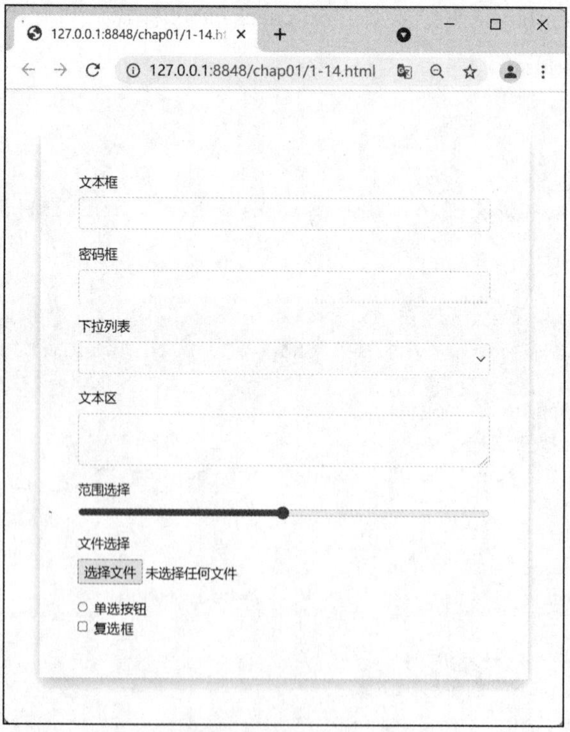

图 1-19 表单控件

1.5.3　表单网格布局

.form-control 会将控件元素宽度设置为 100%，默认布局的控件都是按垂直方向排列的。Bootstrap 允许在表单布局中使用网格来生成更复杂的布局。

【案例 1-16】1-15.html

```html
<!DOCTYPE html>
<html lang="en">
  <head>
    <meta charset="utf-8">
    <meta name="viewport" content="width=device-width, initial-scale=1, shrink-to-fit=no">
    <link rel="stylesheet" href="css/bootstrap.css">
  </head>
  <body>
    <div class="mx-auto my-5 p-5 shadow" style="width: 800px;">
      <form>
        <div class="row">
          <div class="col">
            <div class="form-group">
              <label>文本框</label>
              <input type="text" class="form-control">
            </div>
          </div>
          <div class="col">
            <div class="form-group">
              <label>密码框</label>
              <input type="password" class="form-control">
            </div>
          </div>
          <div class="col">
            <div class="form-group">
              <label>下拉列表</label>
              <select class="form-control">

              </select>
            </div>
          </div>
        </div>
        <div class="row">

          <div class="col-4">
            <div class="form-group">
              <label>文件选择</label>
              <input type="file" class="form-control-file">
            </div>
          </div>
          <div class="col">
            <div class="form-group">
              <label>范围选择</label>
```

```
            <input type="range" class="form-control-range">
          </div>
        </div>
      </div>
      <div class="row">
        <div class="col">
          <div class="form-group">
            <label>文本区</label>
            <textarea class="form-control"></textarea>
          </div>
        </div>
      </div>
    </form>
  </div>
 </body>
</html>
```

代码运行结果如图 1-20 所示。

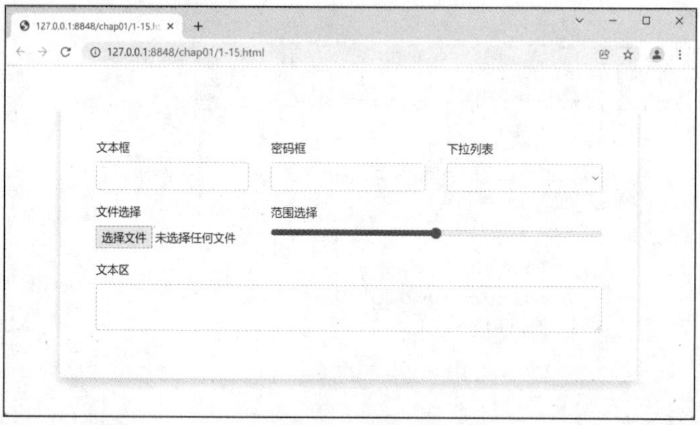

图 1-20 使用网格的表单

Bootstrap 还专门提供了表单使用的网格样式，以简化代码结构。在.form-group 外层加上.form-row 样式的层包裹，同时在.form-group 层自身样式中追加.col 单元格样式就可以实现网格布局。

【案例 1-17】1-16.html

```
<!DOCTYPE html>
<html lang="en">
  <head>
    <meta charset="utf-8">
    <meta name="viewport" content="width=device-width, initial-scale=1, shrink-to-fit=no">
    <link rel="stylesheet" href="css/bootstrap.css">
  </head>
  <body>
    <div class="mx-auto my-5 p-5 shadow" style="width: 800px;">
      <form>
        <div class="form-row">
          <div class="form-group col">
            <label>文本框</label>
```

```html
            <input type="text" class="form-control">
          </div>
          <div class="form-group col">
            <label>密码框</label>
            <input type="password" class="form-control">
          </div>
          <div class="form-group col">
            <label>下拉列表</label>
            <select class="form-control">

            </select>
          </div>
        </div>
        <div class="form-row">
          <div class="form-group col-12">
            <label>文本区</label>
            <textarea class="form-control"></textarea>
          </div>
          <div class="form-group col-12">
            <label>范围选择</label>
            <input type="range" class="form-control-range">
          </div>
        </div>
        <div class="form-row">
          <div class="form-group col-6">
            <label>文件选择</label>
            <input type="file" class="form-control-file">
          </div>
        </div>
      </form>
    </div>
  </body>
</html>
```

代码运行结果如图 1-21 所示。

图 1-21 使用 .form-row 样式的表单

1.5.4 水平表单

如果希望控件标签和控件在水平方向上排列，还是使用网格布局实现。第一步为在.form-group层上加入.row样式，第二步为在控件标签上加入.col-form-label和.col样式，第三步为将控件放入有.col样式的层中。

【案例 1-18】1-17.html

```html
<!DOCTYPE html>
<html lang="en">
  <head>
    <meta charset="utf-8">
    <meta name="viewport" content="width=device-width, initial-scale=1, shrink-to-fit=no">
    <link rel="stylesheet" href="css/bootstrap.css">
  </head>
  <body>
    <div class="mx-auto my-5 p-5 shadow" style="width: 600px;">
      <form>
        <div class="form-group row">
          <label class="col-form-label col-2">文本框</label>
          <div class="col">
            <input type="text" class="form-control">
          </div>
        </div>
        <div class="form-group row">
          <label class="col-form-label col-2">密码框</label>
          <div class="col">
            <input type="password" class="form-control">
          </div>
        </div>
        <div class="form-group row">
          <label class="col-form-label col-2">单选按钮</label>
          <div class="col-10">
            <div class="form-check">
              <input class="form-check-input" type="radio">
              <label class="form-check-label" >
                选项1
              </label>
            </div>
            <div class="form-check">
              <input class="form-check-input" type="radio">
              <label class="form-check-label">
                选项2
              </label>
            </div>
          </div>
        </div>
      </form>
    </div>
  </body>
</html>
```

代码运行结果如图 1-22 所示。

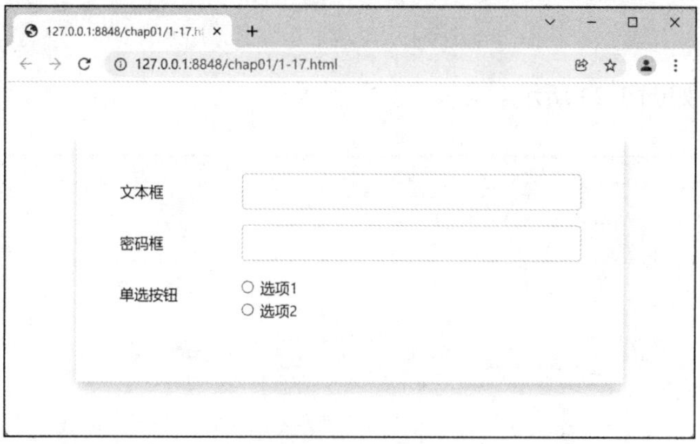

图 1-22 水平表单

1.5.5　行内表单

如果希望所有的控件在一行里显示，可以使用行内表单。行内表单需要在<form>标签上加入.form-inline 样式。

使用.form-inline 样式时，.form-control 样式的元素宽度不再是 100%，而是 auto。

使用.form-inline 样式时，如果内容较多，可能出现一行放不下的情况，这时可以给<label>标签加上.sr-only 样式。.sr-only 样式只在特殊的屏幕阅读器下才显示，一般浏览器不显示。

【案例 1-19】1-18.html

```
<!DOCTYPE html>
<html lang="en">
  <head>
    <meta charset="utf-8">
    <meta name="viewport" content="width=device-width, initial-scale=1, shrink-to-fit=no">
    <link rel="stylesheet" href="css/bootstrap.css">
  </head>
  <body>
    <div class="mx-auto my-5 p-5 shadow" style="width: 800px;">
      <form class="form-inline">
        <label class="sr-only">文本框</label>
        <input type="text" class="form-control mx-2" placeholder="文本框">

        <label class="sr-only">密码框</label>
        <input type="password" class="form-control mx-2" placeholder="密码框">

        <div class="form-check">
          <input type="radio" class="form-check-input" />
          <label class="form-check-label">单选按钮</label>
        </div>
```

```
        </form>
    </div>
  </body>
</html>
```

代码运行结果如图 1-23 所示。

图 1-23　行内表单

1.6　图标

Bootstrap 提供了一套高质量、开源的图标库。使用图标可以美化页面，让内容更直观。

1.6.1　Bootstrap 图标下载

单击 Bootstrap 官网的顶部的"Icons"菜单，打开图标首页，单击下方的"Install"链接，跳转到 Bootstrap 图标安装说明，如图 1-24 所示。

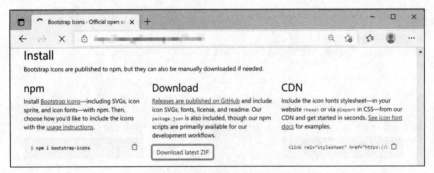

图 1-24　Bootstrap 图标安装

Bootstrap 图标安装有 3 种方式：使用包管理器、下载和 CDN 引入。以下载安装为例，单击"Download latest ZIP"链接跳转到最新的 Bootstrap 图标下载页面，如图 1-25 所示。单击"bootstrap-icons-1.5.0.zip"链接进行下载。

打开下载的压缩包里的 index.html 文件，会显示所有的图标效果及其对应的图标名，在需要查找图标名时可以打开此文件查看。

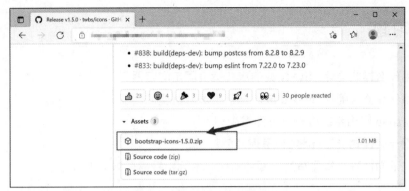

图 1-25　Bootstrap 图标下载

1.6.2　Bootstrap 图标使用

将解压出来的文件夹复制到项目目录，在 HTML 页面中引入 bootstrap-icons.css。要使用 Bootstrap 图标有多种方式，下面的例子分别通过字体图标类和字体 svg 图两种方式使用 Bootstrap 图标。

【案例 1-20】1-19.html

```
<!DOCTYPE html>
<html lang="en">
  <head>
    <meta charset="utf-8">
    <meta name="viewport" content="width=device-width, initial-scale=1, shrink-to-fit=no">

    <link rel="stylesheet" href="css/bootstrap.css">
    <link rel="stylesheet" href="bootstrap-icons-1.5.0/bootstrap-icons.css" />
  </head>
  <body>
    <i class="bi-alarm"></i>
    <i class="bi-alarm text-danger" style="font-size: 2rem;"></i>
    <img src="bootstrap-icons-1.5.0/alarm.svg" alt="alarm">
  </body>
</html>
```

使用字体图标方式时，第一步为在页面中引入图标字体 CSS 文件，第二步为根据需要为 HTML 标签添加对应的 class 名称。名称格式为 bi-{图标名}，例如 class="bi-alarm"。Bootstrap 图标默认大小为 1rem，可以通过修改 font-size 来改变图标大小，通过修改字体颜色来改变图标颜色。

通过图片方式使用时，直接在图片地址中设置对应的 svg 路径，可以通过修改图片的大小改变图标大小。代码运行结果如图 1-26 所示。

图 1-26　Bootstrap 图标效果

本章习题

1. 使用什么样式能让<div>生成响应式宽度固定的容器？
2. 使用什么样式能让<div>生成占据整个屏幕宽度的容器？
3. Bootstrap 网格系统将一行分为多少列？
4. 使用什么样式能让<p>内容呈现红色？
5. 完成下面代码，实现有逐行颜色对比、鼠标指针悬停效果的表格。

```
<table class="_____">
    <tr>
        <td>信息学院</td>
        <td>张三</td>
        <td>zhangsan@example.com</td>
    </tr>
    <tr>
        <td>信息学院</td>
        <td>李四</td>
        <td>lisi@example.com</td>
    </tr>
    <tr>
        <td>计算机学院</td>
        <td>王五</td>
        <td>wangwu@example.com</td>
    </tr>
</table>
```

6. 根据图 1-27 所示效果制作登录表单。

图 1-27　登录表单

第2章
Bootstrap进阶

本章目标

- 掌握 Bootstrap 常用组件
- 掌握 Bootstrap 常用插件

2.1 Bootstrap 组件

Bootstrap 提供了一系列模块化、可重用的组件,例如警告框、按钮和按钮组、输入框组等。组件是页面中独立的代码结构,使用组件可以减少代码耦合,提高开发效率。

2.1.1 警告框

警告框用来显示文本,用法是基础样式.alert 与.alert-primary、.alert-secondary、.alert-success、.alert-danger、.alert-warning、.alert-info、.alert-light 和.alert-dark 共 8 种效果样式配合使用。警告框提供了边框色、背景色和字体颜色的搭配,警告框里面的内容没有特别要求,可以使用其他元素。

【案例 2-1】2-1.html

```html
<!DOCTYPE html>
<html lang="en">
  <head>
    <meta charset="utf-8">
    <meta name="viewport" content="width=device-width, initial-scale=1, shrink-to-fit=no">
    <link rel="stylesheet" href="css/bootstrap.css">
  </head>
  <body>
    <div class="alert alert-primary">
      <h4>操作完成</h4>
      这是一个警告框,如果内容比较少,可以直接写文本
    </div>
    <div class="alert alert-danger">
      <h4>操作失败</h4>
      <p>您的操作因为系统问题失败了</p>
      <p>使用元素可能会导致边距问题</p>
    </div>
    <div class="alert alert-success">
      <h4>操作成功</h4>
      <p class="mb-0">需要时可以通过边距样式修正</p>
    </div>
  </body>
</html>
```

代码运行结果如图 2-1 所示。

使用 Bootstrap 插件可以实现警告框消失效果。实现插件效果需要在代码中按顺序加载 jQuery 和 Bootstrap 的 JS 文件。

第2章 Bootstrap进阶

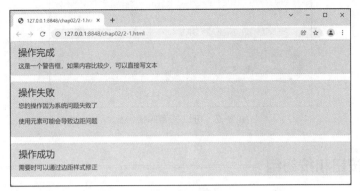

图 2-1 警告框

【案例 2-2】2-2.html

```html
<!DOCTYPE html>
<html lang="en">
  <head>
    <meta charset="utf-8">
    <meta name="viewport" content="width=device-width, initial-scale=1, shrink-to-fit=no">
    <link rel="stylesheet" href="css/bootstrap.css">
  </head>
  <body>
    <div class="alert alert-primary alert-dismissible fade show">
      <button type="button" class="close" data-dismiss="alert">
        <span>&times;</span>
      </button>
      <h4>操作完成</h4>
      <ul class="mb-0">
        <li>
          .fade 设置动画过渡效果
        </li>
        <li>
          .show 设置初始显示
        </li>
        <li>
          .alert-dismissible 设置警告框右侧内边距
        </li>
      </ul>
    </div>
    <script src="js/jquery-3.6.0.min.js"></script>
    <script src="js/bootstrap.bundle.js"></script>
  </body>
</html>
```

代码中的 data-dismiss="alert"属性是一个钩子属性，HTML 元素本来没有这个属性，由 Bootstrap 框架自定义加入，Bootstrap 插件通过读取这个属性知道当前警告框需要监听关闭按钮的单击事件，单击时将警告框隐藏。代码的运行结果如图 2-2 所示。

图 2-2　可以关闭的警告框

2.1.2　按钮和按钮组

Bootstrap 按钮样式可以修改按钮的大小、颜色等。基本用法是基础样式.btn 与.btn-primary、.btn-secondary、.btn-success、.btn-danger、.btn-warning、.btn-info、.btn-light、.btn-dark 和.btn-link 共 9 种效果样式配合使用。

改变按钮大小使用.btn-lg 和.btn-sm。将按钮宽度设置为 100%使用.btn-block。

【案例 2-3】2-3.html

```
<!DOCTYPE html>
<html lang="en">
  <head>
    <meta charset="utf-8">
    <meta name="viewport" content="width=device-width, initial-scale=1, shrink-to-fit=no">
    <link rel="stylesheet" href="css/bootstrap.css">
  </head>
  <body>
    <h4>按钮样式</h4>
    <button type="button" class="btn btn-primary">按钮</button>
    <button type="button" class="btn btn-secondary">按钮</button>
    <button type="button" class="btn btn-success">按钮</button>
    <button type="button" class="btn btn-danger">按钮</button>
    <button type="button" class="btn btn-warning">按钮</button>
    <button type="button" class="btn btn-info btn-lg">大按钮</button>
    <button type="button" class="btn btn-dark btn-sm">小按钮</button>
    <button type="button" class="btn btn-light">按钮</button>
    <button type="button" class="btn btn-link">像链接的按钮</button>
    <h4>超链接使用按钮样式</h4>
    <a class="btn btn-primary" href="#">超链接按钮</a>
    <a class="btn btn-secondary" href="#">超链接按钮</a>
    <a class="btn btn-success" href="#">超链接按钮</a>
    <a class="btn btn-danger" href="#">超链接按钮</a>
    <a class="btn btn-warning" href="#">超链接按钮</a>
    <a class="btn btn-info" href="#">超链接按钮</a>
    <a class="btn btn-light" href="#">超链接按钮</a>
    <a class="btn btn-dark" href="#">超链接按钮</a>
```

```
        <h4>宽度为100%的按钮</h4>
        <button type="button" class="btn btn-primary btn-block">button按钮</button>
        <a class="btn btn-secondary btn-block" href="#">超链接按钮</a>
        <input type="button" class="btn btn-success btn-block" value="input按钮"/>
    </body>
</html>
```

代码运行结果如图2-3所示。

图2-3 按钮

将多个按钮放在.btn-group元素里面构成按钮组，使用按钮组样式可以消除按钮间的间距、修改相邻按钮的圆角，按钮组样式配合Bootstrap插件可以实现更多增强效果。

如果想统一修改按钮组中按钮的大小，可以在.btn-group元素上使用.btn-group-lg和.btn-group-sm。

按钮组里面除了可以放按钮，也可以放入其他的按钮组，例如下拉菜单插件就是通过按钮组嵌套实现的。

【案例2-4】2-4.html

```
<!DOCTYPE html>
<html lang="en">
    <head>
        <meta charset="utf-8">
        <meta name="viewport" content="width=device-width, initial-scale=1, shrink-to-fit=no">
        <link rel="stylesheet" href="css/bootstrap.css">
    </head>
    <body>
        <h4>普通按钮组</h4>
        <div class="btn-group">
            <button class="btn btn-primary">按钮组1按钮</button>
            <button class="btn btn-success">按钮组1按钮</button>
            <button class="btn btn-danger">按钮组1按钮</button>
        </div>
        <h4>大号按钮组</h4>
        <div class="btn-group btn-group-lg ">
            <a class="btn btn-primary">按钮组2按钮</a>
```

```
            <a class="btn btn-success">按钮组2按钮</a>
            <a class="btn btn-danger">按钮组2按钮</a>
        </div>
        <h4>按钮组嵌套</h4>
        <div class="btn-group">
            <button type="button" class="btn btn-secondary">分类管理</button>
            <button type="button" class="btn btn-secondary">商品管理</button>
            <div class="btn-group">
                <button type="button" class="btn btn-secondary dropdown-toggle" data-toggle="dropdown">
                    个人中心
                </button>
                <div class="dropdown-menu">
                    <a class="dropdown-item" href="#">查看</a>
                    <a class="dropdown-item" href="#">退出</a>
                </div>
            </div>
        </div>
        <script src="js/jquery-3.6.0.min.js"></script>
        <script src="js/bootstrap.bundle.js"></script>
    </body>
</html>
```

下拉菜单通过一个按钮组将触发下拉事件的按钮.dropdown-toggle 和下拉菜单项按钮.dropdown-menu 组合在一起。初始状态下外层按钮组只显示内层按钮组里面的.dropdown-toggle 按钮。代码运行结果如图2-4所示。

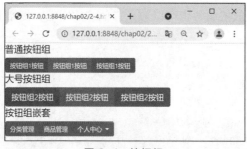

图2-4 按钮组

2.1.3 输入框组

输入框组通过在原生文本控件前后加上文字、按钮等效果来增强效果。输入框组在使用时需要注意.input-group、.input-group-prepend 和.input-group-append 等相关样式的代码结构。

【案例2-5】2-5.html
```
<!DOCTYPE html>
<html lang="en">
    <head>
        <!-- Required meta tags -->
```

```html
    <meta charset="utf-8">
    <meta name="viewport" content="width=device-width, initial-scale=1, shrink-to-fit=no">
    <link rel="stylesheet" href="css/bootstrap.css">
    <link rel="stylesheet" href="bootstrap-icons-1.5.0/bootstrap-icons.css" />
  </head>
  <body>
    <form class="p-4">
      <div class="input-group mb-3">
        <div class="input-group-prepend">
          <p class="input-group-text">
            价格
          </p>
        </div>
        <input type="text" class="form-control" />
      </div>
      <div class="input-group mb-3">
        <input type="text" class="form-control" />
        <div class="input-group-append">
          <p class="input-group-text">
            元
          </p>
        </div>
      </div>
      <div class="input-group mb-3">
        <div class="input-group-prepend">
          <p class="input-group-text">
            <i class="bi bi-currency-yen"></i>
          </p>
        </div>
        <input type="text" class="form-control" />
        <div class="input-group-append">
          <p class="input-group-text">
            元
          </p>
        </div>
      </div>
      <div class="input-group mb-3">
        <input type="text" class="form-control">
        <div class="input-group-append">
          <button class="btn btn-success" type="button">
            <i class="bi bi-search mr-2"></i>搜索
          </button>
        </div>
      </div>
    </form>
  </body>
</html>
```

代码运行结果如图 2-5 所示。

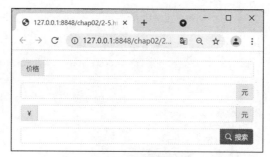

图 2-5　输入框组

2.1.4　列表组

列表组是一种灵活的组件，常用来显示一组相似的内容，例如功能列表、消息列表等。列表组由外层的 .list-group 元素和内层的 .list-group-item 元素构成。

常用的列表组元素可以使用列表、超链接和按钮，也可以根据需要自己组合元素。

【案例 2-6】2-6.html

```
<!DOCTYPE html>
<html lang="en">
  <head>
    <meta charset="utf-8">
    <meta name="viewport" content="width=device-width, initial-scale=1, shrink-to-fit=no">
    <link rel="stylesheet" href="css/bootstrap.css">
    <link rel="stylesheet" href="bootstrap-icons-1.5.0/bootstrap-icons.css" />
  </head>
  <body>
    <ul class="list-group d-inline-flex" style="width: 200px;">
      <li class="list-group-item active">无序列表组</li>
      <li class="list-group-item">无序列表组</li>
      <li class="list-group-item">无序列表组</li>
    </ul>
    <div class="list-group d-inline-flex" style="width: 200px;">
      <a href="#" class="list-group-item list-group-item-action active">
        超链接列表组
      </a>
      <a href="#" class="list-group-item list-group-item-action">
        超链接列表组
      </a>
      <a href="#" class="list-group-item list-group-item-action">
        超链接列表组
      </a>
    </div>
    <div class="list-group d-inline-flex" style="width: 200px;">
      <button type="button" class="list-group-item list-group-item-action active">
        按钮列表组
      </button>
```

```
            <button type="button" class="list-group-item list-group-item-action">
按钮列表组</button>
            <button type="button" class="list-group-item list-group-item-action">
按钮列表组</button>
        </div>
    </body>
</html>
```

使用超链接列表组和按钮列表组时，.list-group-item-action 样式可以修复字体颜色和文字对齐方式。在列表项.list-group-item 元素上使用.active 可以高亮显示当前列表项。代码运行结果如图 2-6 所示。

图 2-6 列表组

2.1.5 分页

项目中经常使用一系列带有数字的链接或者按钮来提醒用户还有未显示的数据，这就是分页。Bootstrap 使用列表来实现分页，标准用法是在<ul>标签上使用.pagination 样式，在<li>标签上使用.page-item 样式，在<a>标签上使用.page-link 样式。

在<ul>上使用.pagination-lg 或者.pagination-sm 可以设置分页组件大小。

在<ul>上使用.justify-content-center 可以实现分页居中对齐，使用.justify-content-end 可以实现分页向右侧对齐。

在<li>上使用.active 或者.disabled 可以设置分页项禁用或者突出显示。

【案例 2-7】2-7.html

```
<!DOCTYPE html>
<html lang="en">
    <head>
        <meta charset="utf-8">
        <meta name="viewport" content="width=device-width, initial-scale=1, shrink-to-fit=no">
        <link rel="stylesheet" href="css/bootstrap.css">
        <link rel="stylesheet" href="bootstrap-icons-1.5.0/bootstrap-icons.css" />
    </head>
    <body>
        <h4>标准分页</h4>
        <ul class="pagination ml-2">
            <li class="page-item active">
                <a class="page-link">&laquo;</a>
            </li>
            <li class="page-item">
```

```html
      <a class="page-link">1</a>
    </li>
    <li class="page-item">
      <a class="page-link">2</a>
    </li>
    <li class="page-item">
      <a class="page-link">3</a>
    </li>
    <li class="page-item">
      <a class="page-link">&raquo;</a>
    </li>
  </ul>
  <h4>大号、居中分页</h4>
  <ul class="pagination justify-content-center pagination-lg">
    <li class="page-item disabled">
      <a class="page-link">&laquo;首页</a>
    </li>
    <li class="page-item active">
      <a class="page-link">1</a>
    </li>
    <li class="page-item">
      <a class="page-link">2</a>
    </li>
    <li class="page-item">
      <a class="page-link">3</a>
    </li>
    <li class="page-item">
      <a class="page-link">&raquo;尾页</a>
    </li>
  </ul>
  <h4>向右对齐分页</h4>
  <ul class="pagination justify-content-end mr-2">
    <li class="page-item">
      <a class="page-link">首页</a>
    </li>
    <li class="page-item">
      <a class="page-link">1</a>
    </li>
    <li class="page-item">
      <a class="page-link">2</a>
    </li>
    <li class="page-item">
      <a class="page-link">3</a>
    </li>
    <li class="page-item">
      <a class="page-link">尾页</a>
    </li>
  </ul>
  </body>
</html>
```

代码运行结果如图 2-7 所示。

图 2-7 分页

2.1.6 巨幕

巨幕组件用来突出显示页面的关键信息。对元素使用.jumbotron 样式，以设置更大的内边距、灰色背景等。可以在巨幕组件里面根据需要放置各种标签。

【案例 2-8】2-8.html

```
<!DOCTYPE html>
<html lang="en">
  <head>
    <meta charset="utf-8">
    <meta name="viewport" content="width=device-width, initial-scale=1, shrink-to-fit=no">
    <link rel="stylesheet" href="css/bootstrap.css">
  </head>
  <body>
    <div class="jumbotron">
      <h1 class="display-4">超市管理系统</h1>
      <p class="lead">使用 Bootstrap4、jQuery、Servlet、JSTL 实现简单的超市进销存管理系统。</p>
      <hr class="my-4">
      <p>系统混合使用了同步和异步请求。为了提高编辑信息时的操作便利性，编辑前数据采用异步方式获取。</p>
    </div>
  </body>
</html>
```

.display-{1-4} 样式设置字体大小、字体加粗程度和行高。.lead 样式设置字体大小和字体加粗程度。代码运行结果如图 2-8 所示。

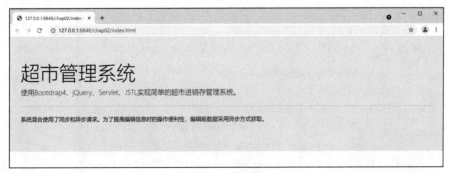

图 2-8 巨幕

2.2 JavaScript 插件

Bootstrap 提供了一系列 JavaScript 插件，JavaScript 插件能够以很简单的方式实现前端交互效果，提高用户体验。Bootstrap 插件的交互效果需要通过 data-*属性或者 jQuery 语法触发，绝大多数情况下使用属性触发就行了，本章选取实战篇需要用到的插件进行讲解。

2.2.1 下拉菜单

下拉菜单插件可以通过单击按钮或者链接来实现内容的显示与隐藏。下拉菜单插件使用了第三方的 popper.js 插件，这个插件已经集成到 bootstrap.bundle.js 和 bootstrap.bundle.min.js 里面。如果使用的是 bootstrap.js，则需要另外引入 popper.js。

下拉菜单的用法如下。

（1）.dropdown 层包裹触发元素和菜单元素。下拉菜单默认向下弹出，在这个层上可以使用.dropup、.dropright 和.dropleft 来改变菜单的弹出方向。

（2）触发元素中指定属性 data-toggle="dropdown"，添加.dropdown-toggle 样式。

Bootstrap 框架会检索 data-toggle 属性值为 toggle 的元素，并为其单击事件绑定下拉菜单行为。

.dropdown-toggle 样式会在触发元素上追加一个提示箭头。

（3）菜单元素样式设置使用.dropdown-menu。

（4）菜单元素的子元素常用样式包括.dropdown-item 菜单项、.dropdown-divider 分割线、.dropdown-header 标题等。

【案例 2-9】2-9.html

```
<!DOCTYPE html>
<html lang="en">
  <head>
    <meta charset="utf-8">
    <meta name="viewport" content="width=device-width, initial-scale=1, shrink-to-fit=no">
    <link rel="stylesheet" href="css/bootstrap.css">
  </head>
  <body>
    <div class="dropdown dropright">
      <button class="btn btn-secondary dropdown-toggle" type="button" data-toggle="dropdown">
        个人中心
      </button>
      <div class="dropdown-menu">
        <h6 class="dropdown-header">信息设置</h6>
        <a href="#" class="dropdown-item">修改信息</a>
        <a href="#" class="dropdown-item">修改密码</a>
        <div class="dropdown-divider"></div>
```

```
          <a href="#" class="dropdown-item">退出</a>
        </div>
      </div>
      <script src="js/jquery-3.6.0.min.js"></script>
      <script src="js/bootstrap.bundle.js"></script>
    </body>
</html>
```

上面的案例演示了一个向右弹出的下拉菜单。代码运行结果如图 2-9 所示。

图 2-9 下拉菜单

2.2.2 折叠

如果需要隐藏和显示大量的内容，可以使用折叠插件。折叠插件最核心的内容由触发元素和折叠元素构成。触发元素包含 data-toggle="collapse" 和 data-target="#target" 两个属性，如果使用超链接作为触发元素，则使用 href 代替 data-target。折叠元素包含 class="collapse" 以及与触发元素 data-target 值一致的 id。

当多个折叠元素放置在同一个父元素里面，同一时刻只能显示其中一个时，就构成手风琴（Accordion）效果。这时需要在折叠元素上添加 data-parent 属性，值为其父元素的 id。

【案例 2-10】2-10.html

```
<!DOCTYPE html>
<html lang="en">
  <head>
    <meta charset="utf-8">
    <meta name="viewport" content="width=device-width, initial-scale=1, shrink-to-fit=no">
    <link rel="stylesheet" href="css/bootstrap.css">
  </head>
  <body>
    <button data-toggle="collapse" data-target="#demo">折叠</button>
    <div id="demo" class="collapse ">
      这里的内容会被折叠/显示
    </div>
    <div id="accordion" class="mt-5">
      <div class="card">
        <div class="card-header">
          <a class="card-link" data-toggle="collapse" href="#collapseOne">
            手风琴效果1
          </a>
        </div>
```

```html
            <div id="collapseOne" class="collapse" data-parent="#accordion">
              <div class="card-body">
                手风琴效果1
              </div>
            </div>
          </div>
          <div class="card">
            <div class="card-header">
              <a class="collapsed card-link" data-toggle="collapse" href="#collapseTwo">
                手风琴效果2
              </a>
            </div>
            <div id="collapseTwo" class="collapse" data-parent="#accordion">
              <div class="card-body">
                手风琴效果2
              </div>
            </div>
          </div>
          <div class="card">
            <div class="card-header">
              <a class="collapsed card-link" data-toggle="collapse" href="#collapseThree">
                手风琴效果3
              </a>
            </div>
            <div id="collapseThree" class="collapse" data-parent="#accordion">
              <div class="card-body">
                手风琴效果3
              </div>
            </div>
          </div>
        </div>
        <script src="js/jquery-3.6.0.min.js"></script>
        <script src="js/bootstrap.bundle.js"></script>
    </body>
</html>
```

代码运行结果如图 2-10 所示。

图 2-10 折叠

2.2.3 导航栏

导航栏是放置在页面顶部用来导航的组件。Bootstrap 的导航栏可以根据屏幕大小展开或者收缩。

一个标准的导航栏由.navbar 样式和响应样式.navbar-expand-{设备}来实现，设备值可以是 sm、md、lg 和 xl，用于设置什么设备以上需要展开导航栏。

导航栏中一般使用.navbar-brand 样式的超链接来放置网址的名字或者 logo 图片。

导航栏中的导航元素可以使用设置了.navbar-nav 样式的<ul>标签，在<ul>标签里面添加使用.nav-item 样式的<li>作导航项，在其中使用.nav-link 样式的<a>作导航链接。

在导航项<li>上可以使用下拉菜单，具体代码与单独使用下拉菜单相似。<li>添加.dropdown样式作为下拉菜单包裹元素，菜单链接添加.dropdown-toggle 样式作为触发元素，另外添加一个下拉菜单弹出元素。

导航栏中可以使用行内表单来做一些简单的交互界面，例如查询关键字或者输入、提交登录信息。

导航栏可以使用颜色帮助样式.bg-*来设置背景色，使用.navbar-dark 和.navbar-light 样式来设置导航字体颜色。

【案例 2-11】2-11.html

```html
<!DOCTYPE html>
<html lang="en">
  <head>
    <meta charset="utf-8">
    <meta name="viewport" content="width=device-width, initial-scale=1, shrink-to-fit=no">
    <link rel="stylesheet" href="css/bootstrap.css">
  </head>
  <body>
    <nav class="navbar navbar-expand bg-dark navbar-dark">
      <a class="navbar-brand" href="#">第 2 章</a>
      <ul class="navbar-nav">
        <li class="nav-item">
          <a class="nav-link" href="#">导航 1</a>
        </li>
        <li class="nav-item">
          <a class="nav-link" href="#">导航 2</a>
        </li>
        <li class="nav-item">
          <a class="nav-link" href="#">导航 3</a>
        </li>
        <li class="nav-item dropdown">
          <a class="nav-link dropdown-toggle" href="#" data-toggle="dropdown">
            下拉导航
          </a>
          <div class="dropdown-menu">
```

```
                <a class="dropdown-item" href="#">Link 1</a>
                <a class="dropdown-item" href="#">Link 2</a>
                <a class="dropdown-item" href="#">Link 3</a>
            </div>
        </li>
    </ul>
    <form class="form-inline" >
        <input class="form-control mr-sm-2" type="text" placeholder="Search">
        <button class="btn btn-success" type="submit">Search</button>
    </form>
</nav>
<script src="js/jquery-3.6.0.min.js"></script>
<script src="js/bootstrap.bundle.js"></script>
</body>
</html>
```

代码运行结果如图 2-11 所示。

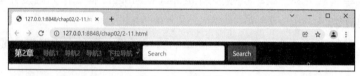

图 2-11　导航栏

有时在一些小屏幕设备上希望导航栏上的一些内容隐藏起来，可以使用折叠导航。

使用一个按钮设置 class="navbar-toggler"、data-toggle="collapse"和 data-target="#thetarget"。然后将需要隐藏的内容放到 class="collapse navbar-collapse"的层中，设置这个层 id 和前面按钮的 data-target 值一致。

下面的案例设置在 lg 宽度以下设备中将导航内容隐藏起来。

【案例 2-12】2-12.html

```
<!DOCTYPE html>
<html lang="en">
  <head>
    <meta charset="utf-8">
    <meta name="viewport" content="width=device-width, initial-scale=1, shrink-to-fit=no">
    <link rel="stylesheet" href="css/bootstrap.css">
  </head>
  <body>
    <nav class="navbar navbar-expand-lg bg-dark navbar-dark">
      <a class="navbar-brand" href="#">第 2 章</a>
      <button class="navbar-toggler" type="button" data-toggle="collapse" data-target="#collapsibleNavbar">
        <span class="navbar-toggler-icon"></span>
      </button>
      <div id="collapsibleNavbar" class="collapse navbar-collapse">
        <ul class="navbar-nav">
          <li class="nav-item">
            <a class="nav-link" href="#">导航 1</a>
          </li>
          <li class="nav-item">
```

```
                <a class="nav-link" href="#">导航 2</a>
            </li>
            <li class="nav-item">
                <a class="nav-link" href="#">导航 3</a>
            </li>
            <li class="nav-item dropdown">
                <a class="nav-link dropdown-toggle" href="#" data-toggle="dropdown">
                    下拉导航
                </a>
                <div class="dropdown-menu">
                    <a class="dropdown-item" href="#">Link 1</a>
                    <a class="dropdown-item" href="#">Link 2</a>
                    <a class="dropdown-item" href="#">Link 3</a>
                </div>
            </li>
        </ul>
        <form class="form-inline">
            <input class="form-control mr-sm-2" type="text" placeholder="Search">
            <button class="btn btn-success" type="submit">Search</button>
        </form>
      </div>
    </nav>
    <script src="js/jquery-3.6.0.min.js"></script>
    <script src="js/bootstrap.bundle.js"></script>
  </body>
</html>
```

2.2.4 模态框

模态框是可以在当前页面上弹出的对话框。其核心内容由触发元素和模态框元素构成。

触发元素包含属性 data-toggle="modal"和 data-target="#myModal"。模态框元素的 id 要和 data-target 的值一致。模态框元素的基本代码结构如下面的案例所示，在模态框上使用.fade 样式添加动画效果，模态框内部含有 data-dismiss="modal"属性的元素可以关闭当前模态框。

【案例 2-13】2-13.html

```
<!DOCTYPE html>
<html lang="en">
  <head>
    <meta charset="utf-8">
    <meta name="viewport" content="width=device-width, initial-scale=1, shrink-to-fit=no">
    <link rel="stylesheet" href="css/bootstrap.css">
  </head>
  <body>
    <!-- 触发模态框 -->
    <button type="button" class="btn btn-primary" data-toggle="modal" data-target="#myModal">
        Open modal
    </button>
```

```html
        <!-- 模态框 -->
        <div class="moda fade1" id="myModal">
          <div class="modal-dialog">
            <div class="modal-content">

              <!-- 模态框 Header -->
              <div class="modal-header">
                <h4 class="modal-title">标题</h4>
                <button type="button" class="close" data-dismiss="modal">&times;</button>
              </div>
              <!-- 模态框 body -->
              <div class="modal-body">
                 正文内容
              </div>
              <!-- 模态框 footer -->
              <div class="modal-footer">
                <button type="button" class="btn btn-danger" data-dismiss="modal">关闭</button>
              </div>
            </div>
          </div>
        </div>
        <script src="js/jquery-3.6.0.min.js"></script>
        <script src="js/bootstrap.bundle.js"></script>
     </body>
</html>
```

代码运行结果如图 2-12 所示。

图 2-12　模态框

本章习题

1. 要创建一个引人注目的容器，可以给<div>添加什么样式？
2. 创建一个大号按钮使用什么样式？
3. 创建按钮组使用什么样式？
4. 创建一个基础的分页使用什么样式？
5. 创建一个基础的导航栏使用什么样式？

6. 根据图 2-13 所示效果完成编码。

图 2-13　习题 6 效果

7. 在习题 6 的基础上实现单击"添加商品"按钮，弹出"添加商品"模态框的效果，如图 2-14 所示。

图 2-14　习题 7 效果

第3章
jQuery基础

本章目标

- 理解 jQuery 框架
- 掌握 jQuery 选择器
- 掌握 jQuery 事件处理
- 掌握 jQuery 操作 DOM 的方法
- 掌握 jQuery AJAX

3.1 jQuery 起步

jQuery 是一种轻量级、功能强大的 JavaScript 框架。与原生 JavaScript 相比，jQuery 提供了统一的应用程序接口（Application Programming Interface，API）进行跨浏览器的元素遍历、生成和操作，能以更简洁的方式实现动画效果，以及进行 AJAX 操作。jQuery 框架是对 JavaScript 的封装，学习本章内容，读者需要掌握 JavaScript 的相关知识。

3.1.1 下载 jQuery

访问 jQuery 官网，单击"Download jQuery"链接跳转到下载页面，如图 3-1 所示。

jQuery 的安装有多种方式：下载 JS 文件、使用包管理器、CDN 引入。这里通过下载 JS 文件的方式进行安装。注意图 3-1 所示页面中有多个下载链接，单击"Download the compressed, production jQuery 3.6.0"链接下载 jquery-3.6.0.min.js，这是压缩后的完整版 jQuery。完整版链接的下方是 slim 版 jQuery 的链接，它精简了 AJAX 模块和特效模块。后面我们需要进行 AJAX 操作，所以要下载完整版的 jQuery。

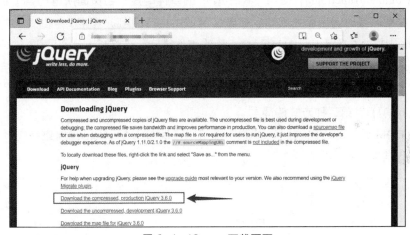

图 3-1　jQuery 下载页面

3.1.2 第一个 jQuery 程序

在 HBuilder X 中使用"基本 HTML 项目"模板新建项目 chap03，将下载的 jquery-3.6.0.min.js 复制到"js"目录。

编辑 index.html，完成后在浏览器中打开，看到页面后打开浏览器的调试模式查看输出。

【案例 3-1】index.html

```
<!DOCTYPE html>
<html>
  <head>
```

```
        <meta charset="utf-8" />
    </head>
<body>
    <p>按【F12】,通过 Chrome 的调试模式查看结果</p>
    <script src="js/jquery-3.6.0.min.js"></script>
    <script>
      window.onload=function(){
        console.log("你好 JavaScript,这是传统 JavaScript 的 onload 事件绑定");
      }
      $(document).ready(
        function(){
          console.log("你好 jQuery,这是标准版 ready 事件绑定");
        }
      );
      $(function(){
        console.log("你好 jQuery,这是简化版 ready 事件绑定");
      });
    </script>
</body>
</html>
```

代码通过 3 种方式在 HTML 文档加载完成后输出一行代码。

window.onload=function(){}使用 JavaScript 的事件绑定。当页面所有内容(包括请求的其他资源)加载完成时调用匿名函数,在匿名函数中执行输出。

$(document).ready(function(){})是 jQuery 的事件绑定。当$(document)通过 jQuery 得到文档对象时,.ready()表示 DOM 加载完成,.ready()中的匿名函数是 DOM 加载事件完成时调用的函数。

$(function(){})是第二种方式的简写。

将操作 DOM 元素的代码放到 ready()里面可以确保 DOM 元素已经创建完成。

代码运行结果如图 3-2 所示。

图 3-2 案例 3-1 运行结果

3.2 jQuery 选择器

3.2.1 jQuery 基本语法

jQuery 的基本语法是 jQuery(selector).action()。jQuery()是一个方法,根据提供的选择器或者 HTML 代码返回匹配的元素集合。.action()是返回的元素所具备的方法。

jQuery()方法的主要用法如下。

（1）jQuery(选择器)：根据 CSS 选择器匹配对应的 DOM 元素，并以 jQuery 对象的形式返回。

（2）jQuery(HTML)：根据提供的 HTML 内容生成 jQuery 对象并返回。

（3）jQuery(object)：将传入的对象包装成 jQuery 对象并返回。将普通对象包装成 jQuery 对象可以增加 jQuery 特性。

（4）jQuery(callback)：当 DOM 元素加载完成时执行方法 callback()。

为了简化代码，jQuery 框架在设计时指定$符号和 jQuery()方法是等价的。上述 jQuery()方法使用说明中的 jQuery 都可以用$符号代替。代码中我们常使用的语法是$(selector).action()。

【案例 3-2】3-1.html

```
<!DOCTYPE html>
<html>
  <head>
    <meta charset="utf-8" />
  </head>
  <body>
    <button id="btn">单击我</button>
    <p id="content"></p>
    <script src="js/jquery-3.6.0.min.js"></script>
    <script>
      $(//step1. start
        function() { //step2. start
          $("#btn").click( //step3. start
            function() { //step4. start
              $("#content").html("Hello jQuery"); //step5
              $("<h4>你好 jQuery</h4>").appendTo("body"); //step6
            } //4. end
          ); //3. end
        } //2. end
      ); //1. end
    </script>
  </body>
</html>
```

代码中 step1 通过$()设置 DOM 元素加载完毕时执行回调方法。

代码中 step2 是 DOM 元素加载完成时需要执行的匿名回调方法 function(){}。

代码中 step3 是在 DOM 元素加载完毕的回调方法中通过选择器找到 id 为"btn"的 DOM 元素，并绑定单击事件。

代码中 step4 是按钮单击事件的匿名回调方法。

代码中 step5 是在按钮单击事件的回调方法中找到 id 为"content"的 DOM 元素，并将内容设置为"Hello jQuery"。

代码中 step6 是在按钮单击事件的回调方法中将"<h4>你好 jQuery</h4>"内容追加到 body 元素中。

代码运行结果如图 3-3 所示，单击按钮后出现两段文字。

图 3-3 jQuery 基本用法示例

3.2.2 jQuery 选择器

通过 jQuery 选择器可以返回 DOM 元素并对其进行操作。jQuery 选择器是基于 CSS 选择器的扩展选择器,它支持 CSS 选择器语法及自定义选择器语法。

常用的 CSS 选择器示例如表 3-1 所示,更多 CSS 选择器请查看 CSS 语法内容。

表 3-1 CSS 选择器的示例

选择器	例子	说明
*	$("*")	通配符选择器
#id	$("#lastname")	id 选择器
.class	$(".intro")	类选择器
element	$("p")	元素标签选择器
:first	$("p:first")	第一个元素
:last	$("p:last")	最后一个元素
:even	$("tr:even")	下标为偶数的元素
:odd	$("tr:odd")	下标为奇数的元素
parent > child	$("div > p")	直接子元素选择器
parent descendant	$("div p")	后代元素选择器
[attribute]	$("[href]")	属性选择器
:first-child	$("p:first-child")	满足条件且同时属于父元素的第一个子元素
:first-of-type	$("p:first-of-type")	父元素中第一个满足条件的子元素
:last-child	$("p:last-child")	满足条件且同时属于父元素的最后一个子元素
:last-of-type	$("p:last-of-type")	父元素中最后一个满足条件的子元素
:input	$(":input")	所有 input 元素
:text	$(":text")	所有类型为 text 的 input 元素
:password	$(":password")	所有类型为 password 的 input 元素
:radio	$(":radio")	所有类型为 radio 的 input 元素
:checkbox	$(":checkbox")	所有类型为 checkbox 的 input 元素
:submit	$(":submit")	所有类型为 submit 的 input 元素
:reset	$(":reset")	所有类型为 reset 的 input 元素
:button	$(":button")	所有类型为 button 的 input 元素
:image	$(":image")	所有类型为 image 的 input 元素
:file	$(":file")	所有类型为 file 的 input 元素
:enabled	$(":enabled")	所有 enabled input 元素
:disabled	$(":disabled")	所有 disabled input 元素
:selected	$(":selected")	所有 selected input 元素
:checked	$(":checked")	所有 checked input 元素

【案例 3-3】3-2.html

```html
<!DOCTYPE html>
<html>
  <head>
    <meta charset="utf-8" />
  </head>
  <body>
    <p>代码中 step1 通过$()设置 DOM 元素加载完毕时执行回调方法。</p>
    <p>代码中 step2 是 DOM 元素加载完成时需要执行的匿名回调方法 function(){}。</p>
    <p>代码中 step3 是在 DOM 元素加载完毕的回调方法中通过选择器找到 id 为"btn"的 DOM 元素，并绑定单击事件。</p>
    <p>代码中 step4 是按钮单击事件的匿名回调方法。</p>
    <p>代码中 step5 是在按钮单击事件的回调方法中找到 id 为"content"的 DOM 元素，并将内容设置为"Hello jQuery"。</p>
    <script src="js/jquery-3.6.0.min.js"></script>
    <script>
      $(function(){
        $("p").css("font-weight","bolder");
        $("p:first-child").css("color","red");
        $("p:odd").css("background-color","gray");
        $("p:odd").css("color","white");
      });
    </script>
  </body>
</html>
```

代码将所有<p>元素通过.css(属性，值)方法设置字体加粗，第一个<p>元素是父元素的第一个子元素，设置为红色字体，所有下标为奇数的<p>被设置为灰色背景、白色字体，由于下标是从 0 开始，所以实际效果是偶数行改变。代码运行结果如图 3-4 所示。

图 3-4　选择器应用示例

3.3　jQuery 集合操作

jQuery 经常需要对集合或其子元素进行遍历和过滤操作，jQuery 提供了相关方法操作集合，如表 3-2 所示。

表 3-2　jQuery 集合常用方法

方法	说明
.add()	添加元素到元素集合
.children()	返回集合中每个元素的子元素
.closest()	从元素本身开始，在 DOM 树中逐级向上匹配，并返回最先匹配的祖先元素
.contents()	返回集合中每个元素的子元素，包括文字和注释节点
.each()	遍历集合，为每个元素执行一个函数
.end()	终止在当前链的最新过滤操作，并返回匹配的元素的以前状态
.eq()	返回集合中指定的元素
.filter()	返回集合中匹配表达式的元素集合
.find()	返回集合中每个符合条件元素的后代
.first()	返回集合中的第一个元素
.has()	返回集合中相匹配的后代元素
.last()	返回集合中的最后一个元素
.map()	通过一个函数匹配当前集合中的每个元素，产生一个对象
.not()	从集合中移除指定的元素
.next()	返回集合中每一个元素紧邻的后面兄弟元素的集合。如果提供一个选择器，那么只有紧跟着的兄弟元素满足选择器的选择条件时，才会返回此元素
.nextAll()	返回集合中每个元素后面的兄弟元素
.nextUntil()	返回集合中每个元素后面的兄弟元素，直到遇到匹配的元素
.offsetParent()	返回离指定元素最近的含有定位信息的祖先元素
.parent()	返回父元素
.parents()	返回所有的祖先元素，可以提供一个可选的选择器作为参数进行筛选
.parentsUntil()	返回当前元素所有的前辈元素，直到遇到匹配的元素
.prev()	返回集合中每一个元素紧邻的前一个兄弟元素的元素集合
.prevAll()	返回集合中每个元素前面的兄弟元素
.prevUntil()	返回集合中每个元素之前的兄弟元素，直到遇到匹配的元素
.siblings()	返回集合中每个元素的兄弟元素
.slice()	根据指定的下标范围，过滤匹配的元素集合，并生成一个新的对象

【案例 3-4】3-3.html

```
<!DOCTYPE html>
<html>
  <head>
    <meta charset="utf-8" />
  </head>
  <body>
    <p>代码中 step1 通过$()设置 DOM 元素加载完毕时执行回调方法。</p>
    <p>代码中 step2 是 DOM 元素加载完成时需要执行的匿名回调方法 function(){}。</p>
```

<p>代码中 step3 是在 DOM 元素加载完毕的回调方法中通过选择器找到 id 为"btn"的 DOM 元素，并绑定单击事件。</p>
<p>代码中 step4 是按钮单击事件的匿名回调方法。</p>
<p>代码中 step5 是在按钮单击事件的回调方法中找到 id 为"content"的 DOM 元素，并将内容设置为"Hello jQuery"。</p>

```html
    <script src="js/jquery-3.6.0.min.js"></script>
    <script>
      $(function(){
        $("body").children().css("border","1px solid");
        $("p").eq(2).css("background-color","yellow");
        $("p").eq(2).next().css("background-color","red");
        $("p").slice(0,1).css("font-size","36px");
        console.log("each()...")
        $("p").each(function(i,e){
          console.log("each",i,e);
        });
        console.log("map()...")
        let obj=$("p").map(function(i,e){
          console.log("map",i,e)
          return $(e).text();
        });
        let obj=$("p").map(function(i,e){
          console.log("map",i,e)
          return $(e).text();
        });
        console.log("obj 是一个数组吗？ ",obj instanceof Array );
        obj=obj.get();
        console.log("obj 是一个数组吗？ ",obj instanceof Array );
        console.log("map return",obj);
      });
    </script>
  </body>
</html>
```

代码通过$("body").children()方法找到 body 所有的子元素，设置 1px 的边框。

代码通过$("p").eq(2)找到下标为 2 的<p>并设置背景为黄色，下标从 0 开始，所以是第三个<p>。

代码通过$("p").eq(2).next()找到下标为 2 的<p>的下一个兄弟元素并设置背景为红色。

代码通过$("p").slice(0,1)找到下标从 0 开始、到 1 结束的<p>，设置字体大小为 36px。

代码通过$("p").each()遍历所有的<p>，匿名函数中第一个参数是下标，第二个参数是元素。

代码通过$("p").map()遍历所有的<p>，并根据需要返回新的元素。匿名函数中第一个参数是下标，第二个参数是元素。map()的匿名方法要返回处理的结果，map()得到的是一个类数组，并不是一个真正的数组（instanceof Array 的结果为 false），可通过 get()方法使其成为真正的数组。

代码运行结果如图 3-5 所示。

图 3-5　集合操作调试模式输出

3.4　jQuery 事件处理

jQuery 提供了一系列方法用于注册事件,以便在用户与浏览器交互时能够监听这些事件,并对事件进行进一步处理。jQuery 事件处理方法如表 3-3 所示。

表 3-3　jQuery 事件处理方法

方法	说明
.off()	移除通过 on 绑定的事件监听
.on()	绑定事件监听
.once()	绑定事件监听,该方法只触发一次
.trigger()	触发元素上的事件及默认行为,允许事件冒泡
.triggerHandler()	触发元素上的事件,但不触发默认行为,阻止事件冒泡,支持自定义事件

下面代码分别为 3 个 <button> 绑定 click 事件处理。第一个按钮被单击时将 <span> 里面的值取到后转换成整数,加 1 后放回 <span>。第二个按钮被单击时将 <span> 里面的值取到后转换成整数,减 1 后放回 <span>。第三个按钮被单击时将所有按钮的 click 事件绑定移除,此后再单击任何按钮都没有变化。

【案例 3-5】3-4.html

```
<!DOCTYPE html>
<html>
  <head>
    <meta charset="utf-8" />
  </head>
  <body>
    <button>+</button><button>-</button><button>off</button>
    <p>
      计数结果为:<span>0</span>
    </p>
    <script src="js/jquery-3.6.0.min.js"></script>
    <script>
```

```
        $(function(){
            $("button").first().on("click",function(){
              let n=parseInt($("span").text());
              $("span").text(n+1);
            });
            $("button").eq(1).on("click",function(){
              let n=parseInt($("span").text());
              $("span").text(n-1);
            });
            $("button").last().on("click",function(){
              $("button").off("click");
            });
        });
    </script>
  </body>
</html>
```

代码运行结果如图 3-6 所示。

下面代码为<a>、<span>和两个<button>绑定 click 事件处理。<a>的事件处理输出 click link，<span>的事件处理输出 click span。第一个按钮被单击时通过.trigger()触发 a span 的 click 事件。第二个按钮被单击时通过.triggerHandler()触发 a span 的 click 事件。

图 3-6　事件处理应用示例

【案例 3-6】3-5.html

```
<!DOCTYPE html>
<html>
  <head>
    <meta charset="utf-8" />
  </head>
  <body>
    <button>trigger</button><button>triggerHandler</button>
    <a href="http://www.baidu.com"><span>百度</span></a>
    <script src="js/jquery-3.6.0.min.js"></script>
    <script>
      $(function(){
        $("button").first().on("click",function(){
          $("a span").trigger("click");
        });
        $("button").last().on("click",function(){
          $("a span").triggerHandler("click");
        });
        $("a").on("click",function(){
          console.log("click link");
        });
        $("span").on("click",function(){
          console.log("click span");
        });
      });
    </script>
  </body>
</html>
```

代码运行后单击第一个按钮可以在调试模式下看到依次输出 click span 和 click link，然后浏览器跳转到百度页面。这说明.trigger()触发了<span>的 click 事件，然后事件冒泡被<a>捕获，最后触发了<a>的默认行为跳转页面。

代码运行后单击第二个按钮可以在调试模式下看到只输出 click span，页面也没有跳转。这说明.triggerHandler()仅触发了<span>的 click 事件，没有事件冒泡。

jQuery 提供了一系列方法来简化事件绑定操作，例如.ready()绑定 DOM 元素加载完成事件，.click()绑定单击事件等。常用事件方法如表 3-4 所示。

表 3-4　jQuery 常用事件方法

方法	说明
.resize()	浏览器事件，调整浏览器大小
.scroll()	浏览器事件，浏览器滚动
.ready()	文档事件，文档加载完成
.blur()	表单事件，元素失去焦点，该事件不会冒泡
.change()	表单事件，元素内容发生变化
.focus()	表单事件，元素获得焦点，该事件不会冒泡
.focusin()	表单事件，元素获得焦点，该事件会冒泡
.focusout()	表单事件，元素失去焦点，该事件会冒泡
.select()	表单事件，元素中的文本被选中
.submit()	表单事件，表单提交
.keydown()	键盘事件，按键按下
.keypress()	键盘事件，按键敲击
.keyup()	键盘事件，按键抬起
.click()	鼠标事件，单击鼠标左键
.contextmenu()	鼠标事件，单击鼠标右键
.dbclick()	鼠标事件，双击鼠标左键
.hover()	鼠标事件，鼠标指针悬停
.mousedown()	鼠标事件，鼠标按键按下
.mouseenter()	鼠标事件，鼠标指针进入元素，该事件不会冒泡
.mouseleave()	鼠标事件，鼠标指针离开元素，该事件不会冒泡
.mousemove()	鼠标事件，鼠标移动
.mouseout()	鼠标事件，鼠标指针离开元素，该事件会冒泡
.mouseover()	鼠标事件，鼠标指针进入元素，该事件会冒泡
.mouseup()	鼠标事件，鼠标按键抬起

下面的例子为含有 required 属性的 input 控件绑定 blur 事件。当事件触发时，判断当前控件的 value 是否有输入，没有输入就在其后的 span 中提示，如果 value 有输入，则将提示内容设置为空。

【案例 3-7】3-6.html

```html
<!DOCTYPE html>
<html>
  <head>
    <meta charset="utf-8" />
  </head>
  <body>
    <form>
      <input type="text" required="required" value=""/><span></span><br>
      <input type="password" required="required" value=""/><span></span><br>
      <button type="submit">提交</button>
    </form>
    <script src="js/jquery-3.6.0.min.js"></script>
    <script>
      $(function(){
        $("input[required='required']").blur(function(){
          let t=$(this).val();
          if(t.length==0){
            $(this).next().text("请输入内容");
          }else{
            $(this).next().text("");
          }
        });
      });
    </script>
  </body>
</html>
```

代码运行结果如图 3-7 所示。

图 3-7　事件绑定应用示例

3.5　jQuery 操作 DOM

jQuery 提供了一系列方法来对 DOM 进行操作，通过这些方法可以修改元素的属性，设置元素的样式，还可以对元素进行插入、复制、删除等操作。常用的方法如表 3-5 所示。

表 3-5　jQuery 常用操作 DOM 的方法

方法	说明
.addClass()	给元素添加一个或者多个类样式，如果是多个样式，则用数组表示
.after()	在元素后面插入参数指定的内容，格式为 target.after(content)
.append()	在元素末尾追加参数指定的内容，格式为 target.append(content)
.appendTo()	将内容追加到参数指定元素的末尾，格式为 content.appendTo(target)

续表

方法	说明
.attr()	获取元素的属性值.attr(key)，或为元素设置属性值.attr(key,value)，如果页面中没有声明属性，则无法取到值
.before()	在元素前面插入参数指定的内容，格式为 target.before(content)
.css()	获取元素的样式属性值，或为每个匹配元素设置一个或多个 CSS 属性。如果一次设置多个 CSS 属性，则使用 JSON 格式数据传值
.html()	以 HTML 格式返回元素内容或者设置元素的 HTML 内容
.insertAfter()	将内容插入参数指定元素的后面，格式为 content.insertAfter(target)
.insertBefore()	将内容插入参数指定元素的前面，格式为 content.insertBefore(target)
.prepend()	在元素开头追加参数指定的内容，格式为 target.prepend(content)
.prependTo()	将内容追加到参数指定元素的开头，格式为 content.prependTo(target)
.prop()	获取元素的属性值.prop(key)，或为元素设置属性值.prop(key,value)，从元素的属性对象中取值，不需要在页面中声明属性
.removeClass()	从元素中移除一个或者多个类样式
.text()	返回元素文本内容或设置文本内容，如果设置内容有标签，会被转义输出
.val()	获取元素的当前 value 值或设置元素的 value 值

下面的例子将 body 的 html 内容转义输出到新建的<p>，并将这个<p>追加到指定的<div>的末尾。通过.addClass()方法为元素应用类样式.hot，通过.css()方法为元素添加两个 CSS 类。

【案例 3-8】3-7.html

```
<!DOCTYPE html>
<html>
  <head>
    <meta charset="utf-8" />
    <style>
      .hot{
        font-size: 24px;
        color: darkred;
        background-color: yellowgreen;
      }
    </style>
  </head>
  <body>
    <div>
      <p>代码中 step1 通过$()设置 DOM 元素加载完毕时执行回调方法。</p>
      <p>代码中 step2 是 DOM 元素加载完成时需要执行的匿名回调方法 function(){}。</p>
      <p>代码中 step3 是在 DOM 元素加载完毕的回调方法中通过选择器找到 id 为"btn"的 DOM 元素，并绑定单击事件。</p>
      <p>代码中 step4 是按钮单击事件的匿名回调方法。</p>
      <p>代码中 step5 是在按钮单击事件的回调方法中找到 id 为"content"的 DOM 元素，并将内容设置为"Hello jQuery"。</p>
    </div>
    <div>
      <h4>body 的内容是</h4>
    </div>
```

```
        <script src="js/jquery-3.6.0.min.js"></script>
        <script>
            $(function(){
                let body_html=$("body").html();
                let div_show=$("div").last();
                let p_content=$("<p></p>");
                p_content.text(body_html);
                div_show.append(p_content);
                $("p").slice(0,2).addClass("hot");
                $("p").odd().css({"border":"1px solid","padding":"15px"});
            });
        </script>
    </body>
</html>
```

代码运行结果如图 3-8 所示。

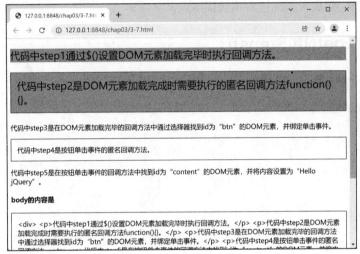

图 3-8　操作 DOM

3.6　jQuery AJAX

3.6.1　jQuery AJAX 简介

AJAX 的全称是 Asynchronous JavaScript And XML（异步 JavaScript 和 XML）。使用 AJAX 技术的 Web 应用能够快速地将增量内容更新在用户界面上，而不需要刷新整个页面，这使得程序能够更快地响应用户的操作。

原生 JavaScript 通过 XMLHttpRequest 对象实现 AJAX 请求，jQuery 将相关操作封装成方法，简化 AJAX 调用流程。

经常使用的 AJAX 简写方法有以下几种。

（1）jQuery.get()实现 get 方式异步请求。

语法为 jQuery.get(url [, data] [, success] [, dataType])。

url 为请求发送的地址。

data 为发送的数据，可以是对象，也可以是字符串。

success 是请求成功后的回调函数 Function(PlainObject d, String t, jqXHR j)。

dataType 是服务器返回的数据格式，其值可以是 XML、JSON、SCRIPT、TEXT、HTML。

（2）jQuery.post()实现 post 方式异步请求。语法和 jQuery.get()一致。

（3）jQuery.getJSON()通过 get 方式请求 JSON 格式的数据。

其语法为 jQuery.getJSON(url [, data] [, success])，跟 jQuery.get()相比少了 dataType 参数，因为返回类型是 JSON。

在前后端异步交互时，经常需要发送数据，jQuery 提供了用于转换数据的方法。

（1）jQuery.param()将对象转换成查询字符串。

（2）.serialize()将表单中的控件值转换成查询字符串。

3.6.2 $.get()和$.post()

下面的例子通过$.get()获取 index.html 的内容，并将其转义输出。

【案例 3-9】3-8.html

```html
<!DOCTYPE html>
<html>
  <head>
    <meta charset="utf-8" />
  </head>
  <body>
    <p></p>
    <script src="js/jquery-3.6.0.min.js"></script>
    <script>
      $(function(){
        $.get("index.html",function(data){
          $("p").text(data);
        });
      });
    </script>
  </body>
</html>
```

代码运行结果如图 3-9 所示。

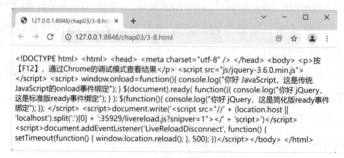

图 3-9　.get()请求应用示例

下面的例子将表单控件的内容通过$.post()异步发送到服务器,由于服务器相关内容还没有讲解,读者可以单击调试模式下的"Network"选项卡查看发送出去的数据。

【案例 3-10】3-9.html

```html
<!DOCTYPE html>
<html>
  <head>
    <meta charset="utf-8" />
  </head>
  <body>
    <form>
      <input type="text"  name="name"/><span></span><br>
      <input type="password" name="pwd"/><span></span><br>
      <button type="button">提交</button>
    </form>

    <script src="js/jquery-3.6.0.min.js"></script>
    <script>
      $(function(){
        $("button").click(function(){
          let form_data=$(this).parent("form").serialize();
          $.post("/login",form_data,function(data){ });
        })
      });
    </script>
  </body>
</html>
```

代码运行结果如图 3-10 所示,输入内容后单击"提交"按钮,单击调试模式下的"Network"选项卡,单击 URL 会在 URL 右侧显示请求响应数据,单击"Headers"选项卡,在最下方可以看到发送出去的数据。

图 3-10　.post()请求应用示例

3.6.3 $.getJSON()

JSON 是一种轻量级的文本格式，具有跨编程语言的特点。JSON 的基本语法如下。

（1）以名值对表示属性和值，名值对之间以冒号分隔。JSON 的值可以是数字、布尔值、文本、对象、数组或者 null。

（2）多个名值对之间以逗号分隔。

（3）以{}表示对象。

（4）以[]表示数组。

下面的例子通过.getJSON()请求 userlist.json 文件，根据返回的 JSON 数据生成列表项。

【案例 3-11】userlist.json

```
{
    "users":[
        {
            "id":1,
            "name":"张三"
        },
        {
            "id":2,
            "name":"李四"
        },
        {
            "id":3,
            "name":"王五"
        }
    ]
}
```

3-10.html

```html
<!DOCTYPE html>
<html>
  <head>
    <meta charset="utf-8" />
  </head>
  <body>
    <button>加载 userlist</button>
    <ul>
    </ul>
    <script src="js/jquery-3.6.0.min.js"></script>
    <script>
      $(function(){
        $("button").click(function(){
          $.getJSON("userlist.json",function(data){
            let userlist=$(data.users);
            userlist.each(function(i,e){
              let li=$("<li></li>");
              li.text(e.id+" - "+e.name);
              $("ul").append(li);
            });
```

```
            });
        });
    });
    </script>
  </body>
</html>
```

代码运行效果如图 3-11 所示。

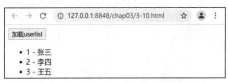

图 3-11 .getJSON()请求应用示例

本章习题

1. jQuery 是基于_____语言的程序库。
2. jQuery 使用_____符号来简写。
3. _____用来选择所有的<div>。
4. _____方法用来设置选中元素的 style 属性。
5. _____可以为所有的<p>设置红色背景。
6. 为了防止 jQuery 代码在文档加载前完成，可以把代码放到_____方法里面。
7. 新建 categories.json 文件，通过 jQuery 实现 AJAX 加载 JSON 文件，并根据内容生成一个下拉菜单，下拉项值为 id，显示内容为 name。

categories.json
```
{
  "data": [
    {
      "description": "分类 1 简介",
      "id": 1,
      "name": "分类 1"
    },
    {
      "description": "分类简介 2",
      "id": 2,
      "name": "分类 2"
    },
    {
      "description": "分类简介 3",
      "id": 3,
      "name": "分类 3"
    }
  ],
  "success": true
}
```

后端篇

第4章
JDBC

本章目标

- 理解 JDBC 基本概念
- 掌握 JDBC 常用 API
- 掌握使用 JDBC 进行 CRUD 操作的方法

4.1 JDBC 基础

在项目开发时对数据进行存储、删除、修改和读取操作，就要用到数据库。Java 程序对数据库进行增、删、改、查操作要用到 JDBC 技术。

学习本章内容前读者需要掌握 Java 基本语法和数据库相关知识，需在计算机上安装好 Java 运行环境。本书的 Java 运行环境是 JDK 1.8 版本。

4.1.1 什么是 JDBC

JDBC 的全称是 Java DataBase Connectivity，它是一组可以发送 SQL 命令到数据库的 API，是由 Java 编写的类和接口。

JDBC 为不同的数据库提供统一的访问方式。各种常见数据库，如 SQL Server、Oracle 和 MySQL，在底层实现上都存在差异，为了通过统一的 API 对这些不同的数据库进行访问，需要这些数据库厂商针对 JDBC 提供的接口实现各自的"驱动类"。

驱动类实现了 JDBC API 中的 java.sql.Driver 接口，该接口用于与数据库服务器进行交互。JDBC 通过这些驱动类操作不同的数据库，实现 Java 程序的通用数据库代码跨数据库运行。JDBC API 与 Java 程序的关系如图 4-1 所示。

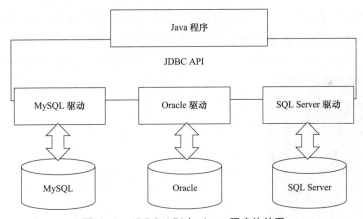

图 4-1　JDBC API 与 Java 程序的关系

4.1.2 环境准备

本书中的数据库代码以 MySQL 8 数据库为例。通过 MySQL 官网打开 MySQL 社区版的下载页面。下载页面中有两个下载链接，体积较小的是在线安装程序，另一个是完整安装程序。单击页面中的"Download"按钮下载安装程序，如图 4-2 所示。

MySQL 8 安装程序集成了很多组件，有一些是我们学习 JDBC 不需要的，在安装时取消相关组件的安装可以降低计算机性能消耗。运行下载的安装程序，在安装类型页面选择"Custom"。

图 4-2　MySQL 社区版下载页面

在组件选择页面只保留组件 MySQL Server 8.x（MySQL 数据库）、MySQL Workbench 8.x（MySQL 图形化管理工具）和 Connector/J 8.x（MySQL 驱动程序），如图 4-3 所示。

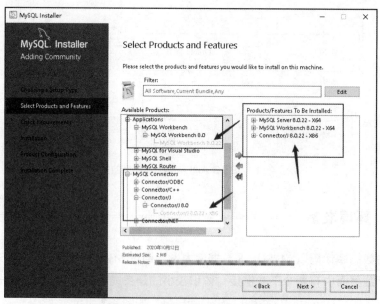

图 4-3　MySQL 安装组件选择

单击"Next"按钮进行环境检查，MySQL 8 需要 MS VC++2019 支持，如果系统中没有安装 MS VC++2019，在这一步会提示，单击"Execute"按钮进行安装。安装好 MS VC++2019 后单击"Next"按钮。

下一步是 MySQL 组件安装，页面会显示将要安装的组件，核对无误后单击"Execute"按钮进行安装，安装结束后结果会在组件右侧显示，如图 4-4 所示。如果安装没有出现问题，单击"Next"按钮开始进行 MySQL 的配置。

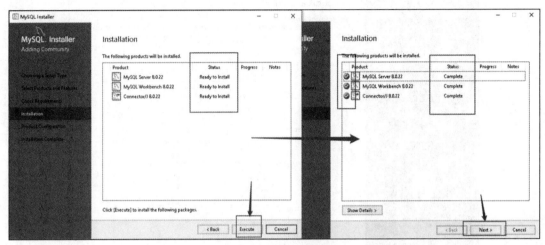

图 4-4　MySQL 组件安装

服务器类型选择"Development Computer"，访问端口保持默认的 3306 端口，如图 4-5 所示。

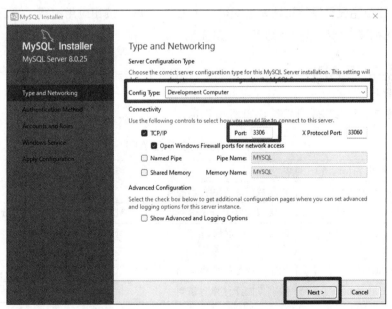

图 4-5　MySQL 网络配置

单击"Next"按钮开始授权设置，这一步需要注意 MySQL 8 默认的密码是强加密模式，提供了更高的安全性，但是会导致一些第三方数据库管理程序或者别的语言（如 PHP）因连接方式不匹配而无法访问数据库。出现这种情况可以通过数据库 alert 命令重新给密码赋值或者在安装 MySQL 8 的时候选择传统授权方式。

接下来设置密码，后面会在程序中使用。本例中密码设置为 123456，如图 4-6 所示。

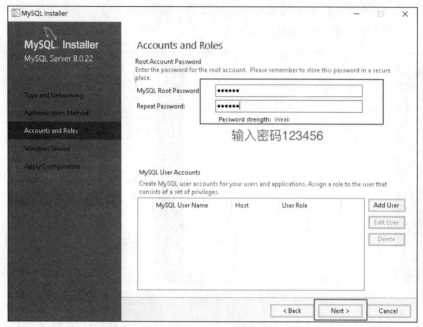

图 4-6　MySQL 密码配置

MySQL 的服务设置保留默认设置即可。单击"Apply"按钮应用配置，最后会出现配置结果，所有的配置结果都是绿色则表示配置完成。

4.1.3　常用 API

JDBC 提供了常用的接口和类，如表 4-1 所示。

表 4-1　JDBC 常用 API

API	说明
java.sql.DriverManager	管理数据库驱动程序的类
java.sql.Connection	连接数据库进行通信的接口，该接口的对象可以与数据库通信
java.sql.PreparedStatement	SQL 语句预处理接口，该接口的对象可以对 SQL 语句进行预处理并发送语句到数据库
java.sql.ResultSet	用来保存数据库查询结果的接口，该接口的对象可以迭代访问查询结果
java.sql.SQLException	数据库操作异常封装类

（1）DriverManager 的常用方法。

static Connection getConnection(String s1,String s2,String s3)静态方法：返回数据库连接对象，参数 s1 是连接字符串，参数 s2 是数据库账号，参数 s3 是数据库访问密码。

（2）Connection 的常用方法。

PreparedStatement prepareStatement(String s)：返回数据预处理对象，参数 s 是需要预处理的 SQL 命令。

void setAutoCommit(boolean b)：设置 JDBC 事务提交方式，参数 b 是事务提交方式。JDBC 事务默认是自动提交的，如果需要手动管理事务，需传入参数值 false。

void commit()：提交事务。如果设置了手动提交事务，需要在预处理代码执行后调用该方法提交事务。

void rollback()：回滚事务。如果设置了手动提交事务，需要在出现异常的时候调用该方法回滚事务。

void close()：释放资源。调用该方法通知数据库关闭和连接有关的资源。

（3）PreparedStatement 的常用方法。

int executeUpdate()：执行会改变数据库的操作，返回值是操作影响的记录数量。

ResultSet executeQuery()：执行查询操作，返回值是查询结果集。

void close()：释放资源。调用该方法通知数据库释放和预处理对象有关的资源。

void setInt(int i ,int v)：为 int 类型占位符赋值，参数 i 是占位符的顺序，参数 v 是需要赋的整数值。

void setString(int i,String s)：为 String 类型占位符赋值，参数 i 是占位符的顺序，参数 s 是需要赋的字符串值。

类似为占位符赋值的方法还有很多，基本格式为 set 数据类型（int i,数据类型 v），其中参数 i 是占位符的顺序，参数 v 是需要赋的值。

（4）ResultSet 的常用方法。

boolean next()：返回结果集中还有没有未访问的记录，如果为 true，则将结果集访问游标指向下一条未访问的记录。

void close()：释放资源。调用该方法通知数据库释放和结果集对象有关的资源。

int getInt(String s)：以 int 类型返回当前记录中字段 s 的值。

String getString(String s)：以 String 类型返回当前记录中字段 s 的值。

类似返回当前记录某个字段值的方法还有很多，基本格式为返回值类型 get 数据类型(String s)，其中参数 s 是当前记录的字段名，如果查询时使用了别名，则 s 的值是别名。

（5）SQL Exception 的常用方法。

void printStackTrace()将错误及其回溯信息通过标准错误流对象输出。

4.1.4　JDBC 操作步骤

JDBC 操作一般分为以下 5 个步骤。

（1）注册驱动。

数据库驱动类需要在程序运行时进行初始化，注意不同数据库的驱动类是不一样的，甚至同一数据库但版本不一样时驱动类也可能是不一样的。

注册驱动使用 Class.forName("驱动类")。

（2）建立数据库连接。

java.sql.Connection 接口的对象负责建立 Java 程序和数据库的资源链接。该对象存在于 JVM 中，和数据库服务器中的连接对象相关联。

建立数据库连接使用 DriverManager.getConnection("连接字符串", "账号", "密码")。

（3）创建 SQL 预处理对象。

JDBC 的 Statement、PreparedStatement 和 CallableStatement 可以让用户发送 SQL 命令到数据库，并从用户的数据库接收数据。

Statement 只能使用静态 SQL 语句，且安全性较差，容易被 SQL 注入威胁。

PreparedStatement 继承 Statement，加入预处理语句的支持，可以使用?符号对 SQL 语句中的值进行占位，随后为?赋值，实现动态 SQL 支持。

CallableStatement 继承 PreparedStatement，添加了对存储过程的支持。

三者都是为了执行 SQL，没有参数的情况下建议使用 Statement，需要使用存储过程时使用 CallableStatement。一般情况下 SQL 语句都需要输入参数，所以 PreparedStatement 使用频率更高，本章以 PreparedStatement 操作为例进行讲解。

获取预处理对象使用 Connection 对象的 prepareStatement("SQL 语句")方法。

得到的预处理对象如果含有占位符，需要通过 setXxx(占位符位置，值)方法为占位符赋值。占位符位置是该占位符出现的位置序号，从 1 开始计数。Xxx 是该占位符对应的数据类型，例如 setInt()、setString()等。

（4）执行 SQL 操作。

SQL 语句的执行分为两种情况：会改变数据库内容的操作（insert 操作、delete 操作、update 操作和 DDL 操作）和不改变数据库内容的操作（select 操作）。

会改变数据库内容的操作通过预处理对象的 executeUpdate()方法执行，返回值 int，表示操作影响的记录数。

不改变数据库的操作通过预处理对象的 executeQuery()方法执行，返回值 ResultSet，表示查询结果，一般需要对迭代 ResultSet 使用查询结果封装 Java 对象。

（5）关闭资源。

Connection、PreparedStatement 和 ResultSet 对象都是存在 JVM 里的，数据库服务器里面会分配相应资源。当它们使用完毕后，需要通过各自的 close()方法通知数据库服务器释放对相数据库资源，而它们自己则会被 JVM 的垃圾回收机制回收。如果不使用 close()方法通知数据库释放资源，程序短时间看起来可以正常执行，但是长时间运行后数据库资源就会耗尽，从而引发程序异常。

4.1.5 第一个 JDBC 程序

本小节使用 IntelliJ IDEA 新建一个 Java 程序，通过 JDBC 新建数据库 chap04。

在 IntelliJ IDEA 中新建项目 chap04，项目类型选择 Java。

在安装 MySQL 时安装了 connector/J，找到 C:\Program Files (x86)\MySQL\Connector J 8.0 目录下的 mysql-connector-java-8.0.xx.jar 文件，这个文件就是 MySQL 8 的 JDBC 驱动。将该文件复制，单击 Intellij IDEA 项目根目录 chap04 后进行粘贴，如图 4-7 所示。

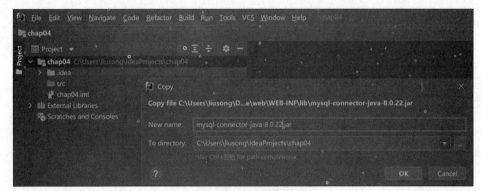

图 4-7　复制驱动

粘贴完成后的驱动是无法在程序运行时找到的，需要先添加到运行库中。选中项目下的驱动文件，单击鼠标右键在弹出的快捷菜单中选择"Add as Library…"命令，在弹出的对话框中单击"OK"按钮，将驱动文件添加到项目的运行库中，如图 4-8 所示。

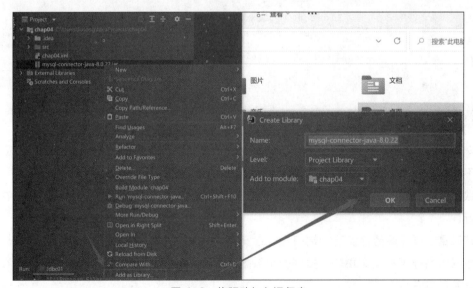

图 4-8　将驱动加入运行库

【案例 4-1】通过 JDBC 创建数据库 chap04

```
import java.sql.Connection;
import java.sql.DriverManager;
import java.sql.PreparedStatement;
import java.sql.SQLException;
```

```java
public class Jdbc01 {
    public static final String JDBC_URL="jdbc:mysql://localhost:3306/?useUnicode=true&characterEncoding=utf-8&serverTimezone=UTC&allowMultiQueries=true";
    public static final String JDBC_USER="root";
    public static final String JDBC_PASSWORD="123456";
    public static void main(String[] args) {
        Connection con=null;
        PreparedStatement ps=null;
        try {
            Class.forName("com.mysql.cj.jdbc.Driver");
            con= DriverManager.getConnection(JDBC_URL,JDBC_USER,JDBC_PASSWORD);
            ps=con.prepareStatement("create database chap04");
            ps.executeUpdate();
            System.out.println("数据库创建完成..");
        } catch (Exception e) {
            e.printStackTrace();
        }finally {
            if(ps!=null){
                try {
                    ps.close();
                } catch (SQLException e) {
                    e.printStackTrace();
                }
            }
            if(con!=null){
                try {
                    con.close();
                } catch (SQLException e) {
                    e.printStackTrace();
                }
            }
        }
    }
}
```

代码中 JDBC_URL 被称为连接字符串，按照下面的结构书写。

协议://数据库服务器 IP 或者域名:数据库监听端口/数据库名?连接参数

不同的数据库协议写法是不一样的，MySQL 的协议是 jdbc:mysql。

数据库服务器 IP 或者域名根据数据库的实际情况书写，localhost 表示数据库安装在本机中运行。

端口是数据库服务器监听的端口号，在安装 MySQL 时默认的端口是 3306。

数据库名标明希望 JDBC 连接到哪个数据库，当前代码还没有创建数据库，所以在此例中数据库名没有给出。

连接参数是 JDBC 传给 MySQL 8 的一些访问设置，例如编码方式、时区设置等。多个设置中间用&符号连接，在 MySQL 8 中缺少某些设置会导致程序无法连接数据库。

JDBC_USER 是数据库连接账号，MySQL 默认账号为 root。

JDBC_PASSWORD 是数据库密码，安装 MySQL 时设置为 123456。

DriverManager.getConnection()方法使用上面的 JDBC_URL、JDBC_USER 和 JDBC_

PASSWORD 3 个参数值创建 Connection 对象。在不同的代码中，3 个参数的内容需要根据实际情况做出改变，例如数据库名、连接密码等。

con.prepareStatement("CREATE DATABASE chap04;")创建了一个代表建库语句的预处理对象。

ps.executeUpdate()通过预处理对象将建库命令发送给数据库。

finally 中按顺序释放预处理对象和连接对象的资源。本例中不涉及查询语句，所以没有使用到 ResultSet 对象。

运行成功后在数据库中会看到刚刚新建的数据库 chap04。

本例演示了通过 JDBC 创建数据库的操作。例如建库、建表等 DDL 命令也可以使用数据库脚本或者数据库图形化管理软件完成。

4.2 JDBC 基本操作

4.2.1 PO 和 DAO

面向数据库 CRUD 的操作主体是表和表中的记录，面向对象增、删、改、查的操作主体是 Java 类和类的对象，我们编写 Java 程序进行数据库操作实际上就是将对类和对象的操作通过 JDBC 转换为对表和记录的操作。

一般称 Java 中和数据库表相对应的类为实体（Entity）类，实体类的属性及其类型和表中的字段及其类型相对应。

实体类对象和表中的记录存在对应关系，称为持久化对象（Persistent Object，PO）。

将 JDBC 操作细节封装成方法供外部调用的类称为数据访问对象（Data Access Object，DAO）。

在 chap04 数据库中新建表 t_user 来存放用户信息，表结构如表 4-2 所示。

表 4-2 t_user 表结构

字段	类型	约束	允许为空	其他	说明
id	int	PK	NO	auto_increment	用户主键
username	varchar(32)		NO		用户名
pwd	varchar(32)		NO		用户密码

在项目 chap04 中新建实体类 po.User，用来和表 t_user 对应，代码如下所示。

```
public class User {
    private int id;
    private String username;
    private String pwd;
    public int getId() {
        return id;
```

```java
    }
    public void setId(int id) {
        this.id = id;
    }
    public String getUsername() {
        return username;
    }
    public void setUsername(String username) {
        this.username = username;
    }
    public String getPwd() {
        return pwd;
    }
    public void setPwd(String pwd) {
        this.pwd = pwd;
    }
    public String toString() {
        return "User{" +
            "id=" + id +
            ", username='" + username + '\'' +
            ", pwd='" + pwd + '\'' +
            '}';
    }
}
```

在项目 chap04 中新建 DAO 类 dao.UserDAO 来封装 JDBC 对表的操作,代码如下所示。

```java
public class UserDAO {
    public static final String JDBC_URL="jdbc:mysql://localhost:3306/chap04?useUnicode=true&characterEncoding=utf-8&serverTimezone=UTC&allowMultiQueries=true";
    public static final String JDBC_USER="root";
    public static final String JDBC_PASSWORD="123456";
    public void save(User user){   }
    public void delete(int id){   }
    public void update(User user){   }
    public User findById(int id){
        return null;
    }
    public List<User> findAll(){
        return null;
    }
}
```

save(User user)将 user 对象的属性值通过 JDBC 添加到表 t_user 中。

delete(int id)根据 id 的值通过 JDBC 从表 t_user 中删除对应主键的记录。

update(User user)将 user 对象的属性值通过 JDBC 更新到表 t_user 中,记录的主键和 user 对象的 id 属性值一致。

findById(int id)根据 id 的值通过 JDBC 从表 t_user 中查询对应主键的记录,并将查询到的记录值封装成 User 对象,返回 User 对象。

findAll()通过 JDBC 查询表 t_user 的所有记录,每一条记录值封装一个 User 对象,并将这些对象放入一个 List 集合,返回 List 集合。

4.2.2 JDBC 添加

【案例 4-2】Jdbc02 调用 UserDAO 的 save()方法完成数据添加

dao.UserDAO 类的 save(Usesr user)方法通过 JDBC 执行 insert 操作，代码如下。

```java
public void save(User user) throws Exception {
    Connection con = null;
    PreparedStatement ps = null;
    try {
        Class.forName("com.mysql.cj.jdbc.Driver");
        con = DriverManager.getConnection(JDBC_URL, JDBC_USER, JDBC_PASSWORD);
        ps = con.prepareStatement("insert into t_user value(null,?,?)");
        ps.setString(1, user.getUsername());
        ps.setString(2, user.getPwd());
        ps.executeUpdate();
    } catch (Exception e) {
        throw new Exception("数据库异常:" + e.getMessage());
    } finally {
        if (ps != null) {
            ps.close();
        }
        if (con != null) {
            con.close();
        }
    }
}
```

DAO 中的方法按照注册驱动、建立数据库连接、创建 SQL 预处理对象、执行 SQL 操作和关闭资源 5 步法完成。

预处理对象执行 insert into t_user value(null,?,?)语句。

value 中第 1 个值对应的是 id 字段，该字段是自增长主键，由数据库自动完成赋值，所以 Java 中给 null。

value 中第 2 个值是?，对应的是 username 字段值，其值是 user 对象的 username 属性值。需要在预处理语句中用?先占位，然后通过 setString(1,user.getUsername())赋值。预处理对象的 setString()表示这是为占位符赋值，值类型是 String；数字 1 表示语句中的第 1 个占位符；user.getUsername()表示 user 对象的 username 属性值。

value 中第 3 个值是?，对应的是 pwd 字段值，其值是 user 对象的 pwd 属性值。需要在预处理语句中用?先占位，然后通过 setString(2,user.getPwd())赋值。预处理对象的 setString()表示这是为占位符赋值，值类型是 String；数字 2 表示语句中的第 2 个占位符；user.getPwd()表示 user 对象的 pwd 属性值。

在 catch 中捕获可能出现的异常，并将异常在方法声明时抛出，这样上层代码在调用 DAO 方法时就需要对异常进行处理。

在 finally 中关闭资源。

在 Jdbc02 中创建一个需要保存到数据库的 User 对象，然后调用 UserDAO 对象的 save()

方法将实体类对象保存到数据库的 t_user 表中,代码如下所示。

```java
public class Jdbc02 {
    public static void main(String[] args) {
        User user=new User();
        user.setUsername("张三");
        user.setPwd("123456");
        UserDAO userDAO=new UserDAO();
        try {
            userDAO.save(user);
        } catch (Exception e) {
            e.printStackTrace();
        }
    }
}
```

执行 Jdbc02,代码运行成功后查看 t_user 表的记录,可以看到新添加的记录。

4.2.3 JDBC 删除

【案例 4-3】Jdbc03 调用 UserDAO 的 delete()方法完成数据删除

dao.UserDAO 类的 delete(int id)方法根据 id 的值通过 JDBC 从表 t_user 中删除对应主键的记录,代码如下所示。

```java
public void delete(int id) throws Exception{
    Connection con = null;
    PreparedStatement ps = null;
    try {
        Class.forName("com.mysql.cj.jdbc.Driver");
        con = DriverManager.getConnection(JDBC_URL, JDBC_USER, JDBC_PASSWORD);
        ps = con.prepareStatement("delete from t_user where id=?");
        ps.setInt(1, id);
        ps.executeUpdate();
    } catch (Exception e) {
        throw new Exception("DAO异常:" + e.getMessage());
    } finally {
        if (ps != null) {
            ps.close();
        }
        if (con != null) {
            con.close();
        }
    }
}
```

预处理对象执行 delete from t_user where id = ? 语句。

删除语句需要提供删除的主键,在执行语句中使用?代替,然后通过 setInt(1,id)赋值。setInt()表示为占位符赋值,值类型是 Int;数字 1 表示语句中的第 1 个占位符;id 表示取参数 id 的值进行赋值。

Jdbc03 中调用 UserDAO 对象的 delete()方法将 id 为 1 的记录从 t_user 表中删除,代码如下所示。

```java
public class Jdbc03 {
    public static void main(String[] args) {
        UserDAO userDAO=new UserDAO();
        try {
            userDAO.delete(1);
        } catch (Exception e) {
            e.printStackTrace();
        }
    }
}
```

代码运行成功后数据库 t_user 表中 id 为 1 的记录就被删除了。

4.2.4 JDBC 修改

【案例 4-4】 Jdbc04 调用 UserDAO 的 update()方法完成数据更新

dao.userDAO 类的 update(User user)方法将 user 对象的属性值通过 JDBC 更新到表 t_user 中，记录的主键和 user 对象的 id 属性值一致。

在运行程序前先向 t_user 表添加一条记录，并记住表中的主键值，代码如下所示。

```java
public void update(User user) throws Exception {
    Connection con = null;
    PreparedStatement ps = null;
    try {
        Class.forName("com.mysql.cj.jdbc.Driver");
        con = DriverManager.getConnection(JDBC_URL, JDBC_USER, JDBC_PASSWORD);
        ps = con.prepareStatement("update t_user set username=? , pwd=? where id=?");
        ps.setString(1, user.getUsername());
        ps.setString(2, user.getPwd());
        ps.setInt(3, user.getId());
        ps.executeUpdate();
    } catch (Exception e) {
        throw new Exception("DAO异常:" + e.getMessage());
    } finally {
        if (ps != null) {
            ps.close();
        }
        if (con != null) {
            con.close();
        }
    }
}
```

预处理对象执行 update t_user set username=? , pwd=? where id=? 语句。

语句中第 1 个?表示需要更新的 username 的值，通过 setString(1,user.getUsername())为它赋值。setString()表示值类型是 String，数字 1 表示是语句中的第 1 个占位符，user.getUsername() 表示用 user 对象的 username 属性值为其赋值。

语句中第 2 个?表示需要更新的 pwd 的值，通过 setString(2,user.getPwd())为它赋值。setString()表示值类型是 String，数字 2 表示是语句中的第 2 个占位符，user.getPwd()表示用

user 对象的 pwd 属性值为其赋值。

语句中第 3 个?表示需要更新的记录的 id 值，通过 setInt(3,user.getId())为它赋值。setInt()表示值类型是 int，数字 3 表示是语句中的第 3 个占位符，user.getId()表示用 user 对象的 id 属性值为其赋值。

在 Jdbc04 中创建一个需要更新到数据库的 User 对象，然后调用 UserDAO 对象的 update()方法将实体类对象更新到数据库的 t_user 表中，代码如下所示。

```java
public class Jdbc04 {
    public static void main(String[] args) {
        User user=new User();
        user.setId(2);
        user.setUsername("丽斯");
        user.setPwd("654321");
        UserDAO userDAO=new UserDAO();
        try {
            userDAO.update(user);
        } catch (Exception e) {
            e.printStackTrace();
        }
    }
}
```

更新成功后数据库 t_user 表中主键为 2 的记录值会发生变化。

4.2.5 JDBC 查询

DAO 的查询操作根据查询结果分为两种情况：结果唯一确定（例如根据主键查询）和结果数量不确定（例如条件查询或者返回所有记录）。结果唯一确定的查询的返回值是实体类，结果数量不确定的查询的返回值一般是集合。

【案例 4-5】 Jdbc05 调用 UserDAO 的 findById()方法和 findAll()方法查询数据

dao.UserDAO 类的 findById(int id)方法根据 id 的值通过 JDBC 从表 t_user 中查询对应主键的记录，结果唯一确定，所以返回值是实体类 User。该方法将查询到的记录值封装成 User 对象，并返回 User 对象，代码如下所示。

```java
public User findById(int id) throws Exception {
    User user = null;
    Connection con = null;
    PreparedStatement ps = null;
    ResultSet rs = null;
    try {
        Class.forName("com.mysql.cj.jdbc.Driver");
        con = DriverManager.getConnection(JDBC_URL, JDBC_USER, JDBC_PASSWORD);
        ps = con.prepareStatement("select * from t_user where id=?");
        ps.setInt(1, id);
        rs = ps.executeQuery();
        if(rs.next()){
            user=new User();
            user.setId(rs.getInt("id"));
            user.setUsername(rs.getString("username"));
```

```
                user.setPwd(rs.getString("pwd"));
            }
        } catch (Exception e) {
            throw new Exception("DAO 异常:" + e.getMessage());
        } finally {
            if(rs!=null){
                rs.close();
            }
            if (ps != null) {
                ps.close();
            }
            if (con != null) {
                con.close();
            }
        }
        return user;
}
```

dao.UserDAO 类的 findAll()方法通过 JDBC 查询表 t_user 的所有记录,结果数量不确定,所以返回值是 List 集合。该方法将每一条记录值封装成一个 User 对象,并将这些对象放入一个 List 集合返回,代码如下所示。

```
public List<User> findAll() throws Exception{
    List<User> userList=new ArrayList<User>();
    Connection con = null;
    PreparedStatement ps = null;
    ResultSet rs = null;
    try {
        Class.forName("com.mysql.cj.jdbc.Driver");
        con = DriverManager.getConnection(JDBC_URL, JDBC_USER, JDBC_PASSWORD);
        ps = con.prepareStatement("select * from t_user");
        rs = ps.executeQuery();
        while(rs.next()){
            User user=new User();
            user.setId(rs.getInt("id"));
            user.setUsername(rs.getString("username"));
            user.setPwd(rs.getString("pwd"));
            userList.add(user);
        }
    } catch (Exception e) {
        throw new Exception("DAO 异常:" + e.getMessage());
    } finally {
        if(rs!=null){
            rs.close();
        }
        if (ps != null) {
            ps.close();
        }
        if (con != null) {
            con.close();
        }
    }
    return userList;
}
```

对比两个查询方法，思考以下内容。

（1）findById()返回唯一记录，所以访问结果集时使用的是 if 判断。findAll()返回记录的数量不确定，所以访问结果集时使用的是 while 判断。

（2）findById()的返回值是实体类，所以返回的 User 在方法开始时声明，初始值为 null，在 if 判断成立后才初始化 User。findAll()的返回值是集合，集合是存放实体类的容器，无论有没有结果容器都是要初始化的，所以返回的集合在方法开始的时候声明并初始化。实体类的声明和初始化都放在 while 判断里面。

Jdbc05 先调用 UserDAO 的 findById()方法查询 id 为 2 的用户信息并显示，然后调用 findAll()方法查询所有用户信息并显示，代码如下所示。

```java
public class Jdbc05 {
    public static void main(String[] args) {
        try {
            UserDAO userDAO = new UserDAO();
            System.out.println("查询 id 为 2 的用户信息");
            User user = userDAO.findById(2);
            System.out.println(user);
            System.out.println("查询所有的用户信息");
            List<User> userList = userDAO.findAll();
            for (User temp : userList) {
                System.out.println(temp);
            }
        } catch (Exception e) {
            e.printStackTrace();
        }
    }
}
```

4.2.6　JDBC 事务

在某些场景里，会一次执行多条能够修改数据库内容的命令，如果在执行过程中发生异常，这些操作需要一起撤销，这时需要把这些操作放在同一个事务中。

JDBC 事务最基本的用法是在开始操作前关闭事务的自动提交功能，在操作完成后手动提交事务，如果出现异常就回滚事务。

【案例 4-6】使用事务回滚操作

Jdbc06 在得到预处理对象前通过 con.setAutoCommit(false)关闭了事务的自动提交功能。在执行删除操作后执行一条有问题的更新操作模拟可能发生的异常。在捕获异常时通过 con.rollback()回滚事务。

```java
public class Jdbc06 {
    public static final String JDBC_URL="jdbc:mysql://localhost:3306/chap04?useUnicode=true&characterEncoding=utf-8&serverTimezone=UTC&allowMultiQueries=true";
    public static final String JDBC_USER="root";
    public static final String JDBC_PASSWORD="123456";
    public static void main(String[] args) {
        Connection con=null;
```

```java
                PreparedStatement ps=null;
                try {
                    Class.forName("com.mysql.cj.jdbc.Driver");
                    con= DriverManager.getConnection(JDBC_URL,JDBC_USER,JDBC_PASSWORD);
                    con.setAutoCommit(false);
                    System.out.println("删除 id 为 2 的用户");
                    ps=con.prepareStatement("delete from t_user where id=2;");
                    ps.executeUpdate();
                    System.out.println("更新 id 为 4 的用户,注意关键字 where 故意写错了,模拟可能发生的异常");
                    ps=con.prepareStatement("update t_user username='haha' whee id=4");
                    ps.executeUpdate();
                    con.commit();
                } catch (Exception e) {
                    if(con!=null){
                        try {
                            con.rollback();
                            System.out.println("事务回滚,删除操作不会发生");
                        } catch (SQLException ex) {
                            ex.printStackTrace();
                        }
                    }
                    e.printStackTrace();
                }finally {
                    if(ps!=null){
                        try {
                            ps.close();
                        } catch (SQLException e) {
                            e.printStackTrace();
                        }
                    }
                    if(con!=null){
                        try {
                            con.close();
                        } catch (SQLException e) {
                            e.printStackTrace();
                        }
                    }
                }
            }
        }
```

代码执行完毕后可以看见文字提示,删除代码执行了,但是同一个事务中的更新操作出现问题,删除操作被一起回滚了,所以数据库中的数据没有发生变化。

本章习题

1. 请简述 JDBC 编程的基本流程。
2. 根据表 4-3 在数据库中新建表。

表 4-3 t_category 表结构

字段	类型	约束	允许为空	其他	说明
id	int	PK	NO	auto_increment	主键
name	varchar(45)		NO		分类名
description	text		YES		分类描述

3. 根据习题 2 新建对应的 po 类。

4. 根据习题 2、习题 3 新建 dao 类，完成对表 t_category 的添加、删除、修改和查询操作。

第5章
Servlet与JSP

本章目标

- 理解什么是 Servlet、JSP
- 掌握常用的 Servlet API
- 掌握 EL 和 JSTL
- 掌握过滤器的使用方法
- 掌握 MVC 模型

5.1 Servlet 基础

广义上的 Servlet 是指用 Java 编写、运行在 Web 服务器中的类，它基于请求响应模型，可以接收 Web 请求并做出响应。狭义上的 Servlet 是指一个 javax.servlet.Servlet 接口，Servlet 接口是所有广义 Servlet 直接或者间接实现的接口。

学习本章内容前读者需要了解 HTTP，掌握 Java 基本语法和 HTML 表单相关知识，并在计算机上安装好 Java 运行环境。本书的 Java 运行环境是 JDK 1.8 版本。

5.1.1 环境准备

Servlet 可以运行在任何支持 Servlet 技术的 Web 服务器中，本书案例以 Tomcat 8 作为运行服务器。

打开 Tomcat 官网的下载页面，选择下载适合操作系统的解压版 Tomcat，如图 5-1 所示。不建议下载 Windows Service installer 安装版，因为该版本会在开机时启动 Tomcat 服务，在开发过程中造成端口冲突。

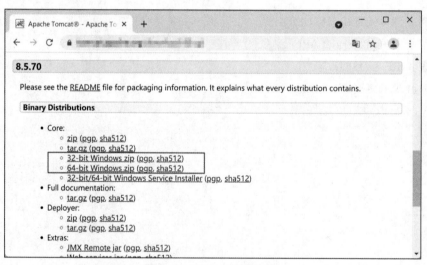

图 5-1　Tomcat 下载页面

将下载的 ZIP 包解压到 C 盘根目录。

启动 Intellij IDEA，选择"File -> Setting"命令打开设置页面。选择左侧导航中的"Application Servers"选项，然后单击中间的 +符号，在弹出的菜单中选择"Tomcat Server"命令，如图 5-2 所示。

在弹出的对话框中单击"Tomcat Home"后的文件选择按钮，选中刚才解压的 Tomcat 的目录，单击"OK"按钮，如图 5-3 所示。如果路径选择没有问题，会正确显示 Tomcat 版本。

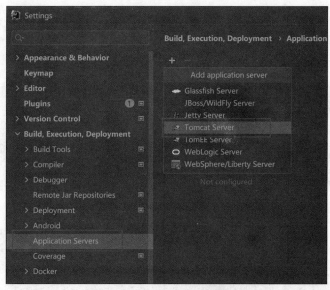

图 5-2 设置 Tomcat 服务器

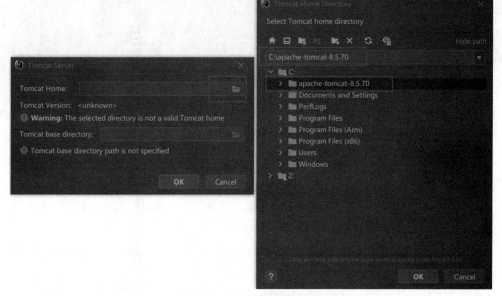

图 5-3 选择 Tomcat 目录

5.1.2 第一个 Web 应用程序

Web 应用程序是从 Web 访问的应用程序。Web 应用程序由 Web 组件（如 Servlet、JSP、过滤器等）和其他元素（如 HTML、CSS 和 JavaScript）组成。

在 Intellij IDEA 中新建 Java 类型项目，项目名为 chap05。选中项目并单击鼠标右键，在弹出的快捷菜单中选择"Add Framework Support…"命令，然后在弹出的内容中勾选"Web Application"，如图 5-4 所示。

选择"File -> Project Structure"命令打开设置页面，选择左侧导航中的"Libraries"选项，然后单击中间的+符号，在弹出的菜单中选择"Java"命令。在弹出的对话框中选中 Tomcat 安装路径 lib 目录下的 servlet-api.jar，单击"OK"按钮，如图 5-5 所示。

图 5-4 添加 Web 支持

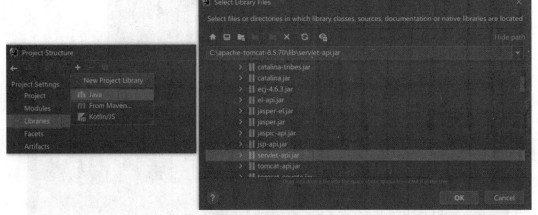

图 5-5 添加 Library

操作完成后项目目录会出现 web 文件夹，External Libraries 下会出现"servlet-api"选项。用同样的方式添加 jsp-api.jar 到 External Libraries 下。

编写的 Java 源代码（如 Servlet、DAO、PO 等）放在 src 目录下各自的包中，JSP、CSS、JavaScript、图片等放到 web 目录下对应的文件夹。

创建一个自定义 Servlet 最简单的做法是继承 javax.servlet.http.HttpServlet 类。想让它处理来自浏览器超链接的请求，就重写 doGet()方法。

【案例 5-1】编写 Servlet 处理超链接请求

在 src 文件夹下新建类 servlet.HelloServlet，代码如下。

```
package servlet;

import javax.servlet.ServletException;
import javax.servlet.annotation.WebServlet;
import javax.servlet.http.HttpServlet;
import javax.servlet.http.HttpServletRequest;
import javax.servlet.http.HttpServletResponse;
import java.io.IOException;
import java.io.PrintWriter;

@WebServlet("/hello")
```

```java
public class HelloServlet extends HttpServlet {
    protected void doGet(HttpServletRequest req, HttpServletResponse resp)
throws ServletException, IOException {
        PrintWriter out =resp.getWriter();
        out.println("hello servlet");
        out.flush();
        out.close();
    }
}
```

代码中@WebServlet("/hello")注解完成了虚拟路径/hello 和 HelloServlet 类的映射绑定。

doGet()方法重写了从 HttpServlet 继承来的方法。它的两个参数 HttpServletRequest req 和 HttpServletResponse resq 分别对应请求对象和响应对象。

PrintWriter 可以向客户端浏览器输出数据，输出的数据内容会被客户端浏览器执行。

编辑 web 文件夹下的 index.jsp，代码如下所示。

```jsp
<%@ page contentType="text/html;charset=UTF-8" language="java" %>
<html>
  <head>
    <title>Hello JSP</title>
  </head>
  <body>
  Hello JSP. <a href="/hello">请求 HelloServlet</a>
  </body>
</html>
```

现在我们有了最简单的 Web 程序，包含一个 JSP 和一个 Servlet，第一次运行还得配置 Web 项目运行服务器和发布路径。单击 Intellij IDEA 工具栏中的"Add Configuration"，在弹出的对话框中单击"Add new run configuration..."链接，然后选择弹出菜单里的"Tomcat Servet->Local"命令，如图 5-6 所示。

解压 Tomcat 后如果正确配置了 Application Server，刚刚操作完成后在左侧会看见 Tomcat 8.x 的选项。切换到"Deployment"选项卡，单击+符号，在弹出的菜单中选择"Artifact"命令将项目发布到 Tomcat 服务器。操作完成后将"Application Context"的内容改成 /符号，如图 5-7 所示。

图 5-6　添加 Tomcat 服务器到项目

操作完成后工具栏中会出现 Tomcat 8.x 的选项，单击绿色三角形图标可以普通模式运行 Web 程序，单击绿色虫子图标可以调试模式运行 Web 程序，如图 5-8 所示。

单击绿色三角形图标，服务器启动完成后会自动打开浏览器，如果能看到文字 Hello JSP 和一个请求 HelloServlet 的超链接，表示 Web 程序在服务器上正常发布并运行，单击超链接，地址栏变成 localhost:8080/hello，页面显示"hello servlet"，表示 Servlet 被正确请求并执行，如图 5-9 所示。

图 5-7 添加部署

图 5-8 运行按钮　　　　　　　　图 5-9 运行 Web 程序

5.1.3　Servlet 运行原理

在前一小节的例子中通过超链接请求 http://localhost:8080/hello 可以运行 HelloServlet 的 doGet()方法并看到运行结果。

http 表示这是一个基于 HTTP 的请求。

localhost 表示 Web 服务器运行在当前主机。

8080 表示 Web 服务器监听的端口是 8080（Tomcat 配置的默认端口是 8080，可以修改）。

/hello 是一个虚拟路径，和 HelloServlet 中的@WebServlet("/hello")里面的值对应。

超链接发送的 HTTP 请求默认是 get 类型。HTTP 请求类型有 get、post、put、delete 等 8 种，其中最常见的是 get 和 post。get 通过 URL 来传递用户的数据，长度有限制且明文可见。post 把数据封装到请求体中发送，相对更安全。

Servlet 运行原理如图 5-10 所示，客户浏览器发送 http://localhost:8080/hello 请求后按以下步骤执行。

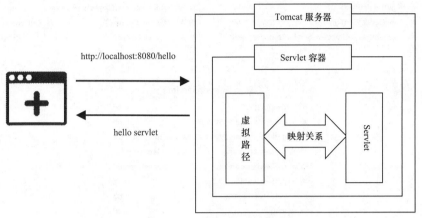

图 5-10　Servlet 运行原理

（1）上述请求根据 http://localhost:8080/，会被发送到本机安装的 Tomcat 服务器。

（2）Tomcat 服务器会获得请求的具体资源虚拟路径/hello，并将这个虚拟路径交给 Tomcat 内部的 Servlet 容器做进一步处理。

（3）Servlet 容器会维护虚拟路径和对应 Servlet 类的映射关系。它在接收到请求/hello 后根据映射关系找到/hello 虚拟路径对应的 HelloServlet 类。

（4）Servlet 容器根据请求类型调用与 Servlet 类匹配的对应的 doXxx()方法，例如当前是 get 类型的 HTTP 请求，就调用 doGet()方法。

（5）HelloServlet 的 doGet()方法执行过程通过 PrintWriter 对象向客户浏览器返回数据，客户浏览器运行返回数据得到运行结果。

5.1.4　Servlet 生命周期

图 5-11 描述了 Servlet 类之间的关系。接口 javax.servlet.Servlet 定义了与 Servlet 生命周期有关的方法。接口 javax.servlet.ServletConfig 定义了获取 Servlet 配置和上下文的方法。

抽象类 javax.servlet.GenericServlet 实现了 Servlet 接口和 ServletConfig 接口中大部分的方法，除了 service()方法。

类 javax.servlet.http.HttpServlet 继承 GenericServlet，实现了 service()方法，在 service()方法中获取请求类型并调用其私有的 doXxx()方法。

GenericServlet 的 service()方法用于处理通用请求和响应，HttpServlet 中添加了对 HTTP 的支持，可以直接访问 HTTP 的请求和响应。

自定义一个能够在 Servlet 容器里运行的 Servlet 有 3 种方式：实现 Servlet 接口、继承 GenericServlet 类并实现 service()方法、继承 HttpServlet 类并重写自己需要的 doXxx()方法。对于常用的 Web 程序，最方便的是继承 HttpServlet 类。

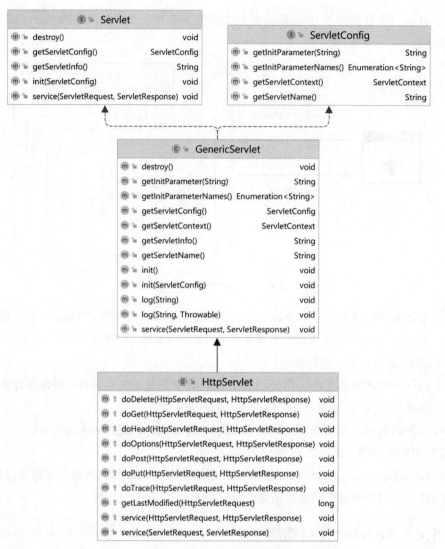

图 5-11　Servlet 类关系图

　　Servlet 生命周期由 Servlet 容器控制，当请求一个 Servlet 实例时，容器按照以下步骤执行。

　　（1）如果 Servlet 实例不存在，那么容器加载 Servlet 类来创建一个实例，调用实例的 init() 方法完成实例初始化。

　　（2）调用 Servlet 实例的 service() 方法，将请求和响应对象作为参数传入。

　　（3）如果需要移除 Servlet 实例，容器会运行 Servlet 实例的 destroy() 方法。

　　init() 方法和 destroy() 方法在 Servlet 生命周期中只运行一次，service() 方法只要请求 Servlet 实例就会运行。

5.1.5　Servlet 请求和响应

　　javax.servlet.ServletRequest 接口的对象用于获取客户端的请求信息，如内容类型、内容长度、参数名称和值、头信息、属性等。接口 javax.servlet.http.HttpServletRequest 继承 ServletRequest，在其基础上增加了一些针对 HTTP 的方法。请求对象常用方法如表 5-1 所示。

表 5-1 请求对象常用方法

方法	说明
String getParameter(String s)	ServletRequest 的方法，从请求中获取参数 s 的值
String[] getParameterValues(String s)	ServletRequest 的方法，以数组方式获取参数 s 的值，主要用于获取复选框的取值
ServletInputStream getInputStream()	ServletRequest 的方法，返回从请求体中读取数据的输入流，主要用来读取上传的文件数据
Object getAttribute(String s)	ServletRequest 的方法，从请求中获取属性 s 对应的值
void setAttribute(String s, Object o)	ServletRequest 的方法，将对象 o 以关键字 s 存入请求的属性中
ServletContext getServletContext()	ServletRequest 的方法，获取上下文 ServletContext 对象
void setCharacterEncoding(String s)	ServletRequest 的方法，设置请求的编码方式为 s
RequestDispatcher getRequestDispatcher(String s)	ServletRequest 的方法，获取路径 s 对应的 RequestDispatcher 对象
String getContextPath()	HttpServletRequest 的方法，获取上下文路径
Cookie[] getCookies()	HttpServletRequest 的方法，以数组的方式获取 Cookie
String getRequestURI()	HttpServletRequest 的方法，获取请求的 URI
String getRequestURL()	HttpServletRequest 的方法，获取请求的 URL
Session getSession(boolean b)	HttpServletRequest 的方法，获取 Session，b 为 true 时，没有 session 就创建新 session 返回

javax.servlet.ServletResponse 接口的对象帮助 Servlet 向客户端响应数据。接口 javax.servlet.http.HttpServletResponse 继承 ServletResponse，在其基础上增加了一些针对 HTTP 的方法。响应对象常用方法如表 5-2 所示。

表 5-2 响应对象常用方法

方法	说明
PrintWrite getWriter()	ServletResponse 的方法，获取 PrintWriter 对象
void setCharacterEncoding(String s)	ServletResponse 的方法，设置响应的编码方式为 s
void setContentType(String s)	ServletResponse 的方法，设置响应类型为 s
void addCookie(Cookie c)	HttpServletResponse 的方法，添加 Cookie 到响应
void sendRedirect(String s)	HttpServletResponse 的方法，重定向到地址 s

【案例 5-2】register.jsp 表单

```
<%@ page contentType="text/html;charset=UTF-8" language="java" %>
<html>
<head>
```

```html
        <title>注册表单</title>
        <style>
        </style>
</head>
<body>
<form action="/demo01" method="post">
    <table>
        <tr>
            <td>姓名</td>
            <td><input type="text" name="username"/></td>
        </tr>
        <tr>
            <td>密码</td>
            <td><input type="password" name="pwd"/></td>
        </tr>
        <tr>
            <td>性别</td>
            <td><input type="radio" name="gender" value="男" checked/>男<input type="radio" name="gender" value="女"/>女</td>
        </tr>
        <tr>
            <td>爱好</td>
            <td>
                <input type="checkbox" name="hobbies" value="football"/>足球
                <input type="checkbox" name="hobbies" value="basketball"/>篮球
                <input type="checkbox" name="hobbies" value="reading"/>阅读
                <input type="checkbox" name="hobbies" value="music"/>音乐
            </td>
        </tr>
        <tr>
            <td>
                <button type="submit">提交</button>
            </td>
        </tr>
    </table>
</form>
</body>
</html>
```

register.jsp 的第一行代码<%@ page contentType="text/html;charset=UTF-8" language="java"%>称为 JSP 的 page 指令标签，设置 JSP 的内容格式为 html，编码方式为 UTF-8。除了第一行代码，剩下的全部是 HTML 代码，JSP 从书写语法上可以看作将 Java 内容通过 JSP 的标签嵌套到 HTML 代码中。

表单里面的 action="/demo01"设置表单内容提交路径是/demo01，这个路径和下面 Demo01Servlet 虚拟路径设置@WebServlet("/demo01")的参数值保持一致。method="post"设置表单提交到服务器的方式是 post，容器会根据 post 提交方式调用 Demo01Servlet 的 doPost()方法。

```java
package servlet;
//省略包导入代码
@WebServlet("/demo01")
public class Demo01Servlet extends HttpServlet {
    protected void doPost(HttpServletRequest req, HttpServletResponse resp) throws ServletException, IOException {
        String url = req.getRequestURL().toString();
        String uri = req.getRequestURI();
        String username = req.getParameter("username");
        String pwd = req.getParameter("pwd");
        String gender = req.getParameter("gender");
        String[] hobbies = req.getParameterValues("hobbies");
        System.out.println("下面的内容会输出到服务器的命令行");
        System.out.println("url = " + url);
        System.out.println("uri = " + uri);
        System.out.println("username = " + username);
        System.out.println("pwd = " + pwd);
        System.out.println("gender = " + gender);
        System.out.print("hobbies = ");
        if (hobbies != null) {
            Arrays.stream(hobbies).forEach(s -> System.out.print(s + " "));
        }
        PrintWriter out = resp.getWriter();
        out.println("下面的内容会输出到客户端浏览器");
        out.println("url = " + url);
        out.println("uri = " + uri);
        out.println("username = " + username);
        out.println("pwd = " + pwd);
        out.println("gender = " + gender);
        out.print("hobbies = ");
        if (hobbies != null) {
            Arrays.stream(hobbies).forEach(s -> out.print(s + " "));
        }
    }
}
```

Demo01Servlet 的 doPost()方法从请求对象获取表单提交的 username、pwd、gender 和 hobbies 参数的值。Hobbies 复选框的值以数组方式获得，在遍历 hobbies 前加入非空判断，避免没有选择爱好导致空指针异常。

Demo01Servlet 分别通过 System.out 和 resp.getWriter()输出内容。System.out 是当前系统的命令行输出对象，它输出的内容在服务器端。resp.getWriter()是向客户端浏览器输出数据的对象，它输出的内容会通过服务器响应发给客户端浏览器，浏览器接收到内容后默认按照网页内容输出。

启动服务器，将浏览器地址改为 http://localhost:8080/demo01，在页面中输入内容并单击"提交"按钮，可以看到浏览器页面变成图 5-12 所示的页面，同时 IDEA 的命令行输出内容如图 5-13 所示。

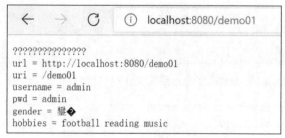

图 5-12 注册表提交后的页面

图 5-13 注册表单提交后服务器命令行输出

5.1.6 乱码处理

前一小节的例子运行成功后发现结果中出现了部分乱码的现象，乱码可以分为输入乱码和输出乱码。

【案例 5-3】处理 Servlet 中文乱码

通过浏览器的调试模式查看注册表单的请求信息，gender 的信息是正常的中文，如图 5-14 所示。这说明 gender 值乱码是数据传入 Servlet 后在取值的过程中出现的，称为输入乱码。

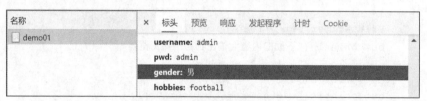

图 5-14 注册表单的请求数据

输入乱码的处理分为以下两步。

（1）在 JSP 页面中通过 page 指令<%@ page contentType="text/html;charset=UTF-8" language="java" %>设置页面的编码方式为 UTF-8。

（2）在 Servlet 中在通过请求对象取参数值之前，通过请求对象调用 setCharacterEncoding("utf-8")方法设置请求的编码方式。

前一小节的例子中第一步已经完成，只需要修改 Demo01Servlet，在 doPost()方法的第一行插入 req.setCharacterEncoding("utf-8"); 语句。

```
protected void doPost(HttpServletRequest req, HttpServletResponse resp) throws ServletException, IOException {
    req.setCharacterEncoding("utf-8");
    //省略其他代码
}
```

重启服务器后执行注册表单并输入内容后提交,IDEA 的命令行 gender 值中文输出已经正常，但是浏览器页面的中文依旧是乱码。这说明中文值在服务器端是正常的，通过 PrintWriter 对象发送给客户端后出现乱码，称为输出乱码。

输出乱码处理需要在获取 PrintWriter 对象前设置响应内容格式和响应编码。通过响应对象调用 setContentType("text/html;charset=utf-8")方法进行设置。

修改 Demo01Servlet，在 doPost()方法的第二行插入 resp.setContentType("text/html;charset=utf-8"); 语句。

```
protected void doPost(HttpServletRequest req, HttpServletResponse resp) throws ServletException, IOException {
    req.setCharacterEncoding("utf-8");
    resp.setContentType("text/html;charset=utf-8");
    //省略其他代码
}
```

重启服务器后执行注册表单并输入内容后提交，这时页面中的中文已经能正常显示。由于设置的页面内容格式为 html，这时输出内容不会自动换行，运行结果如图 5-15 所示。如果想要输出换行，需要在 PrintWriter 输出时以
标签结尾。

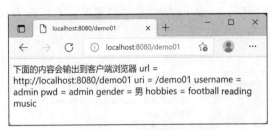

图 5-15　输出中文处理

5.1.7　重定向与转发

大多数时候程序不会直接在 Servlet 中输出大量的 HTML 内容，因为使用 PrintWriter 输出标签很麻烦，如果需要大量输出内容，可以使用 JSP 页面。Servlet 在处理完请求后，可以跳转到下一个资源地址，这个资源可以用来显示数据的 JSP。

Servlet 跳转页面有两种方式：重定向和转发。

重定向通过响应对象的重定向方法向客户端浏览器返回重定向指令和重定向地址，客户端浏览器接收指令后重写请求新地址。

转发通过请求对象获取 RequestDispatcher 转发对象，然后通过该对象的 forward()方法请求新地址，由新地址返回响应结果。

重定向和转发的区别。

（1）重定向是两次请求，第一次由用户主动发起，第二次由浏览器根据响应指令自动发起，重定向后和重定向前的请求对象不同。转发是一次请求，同一个请求在服务器内部转发给不同资源。

（2）重定向可以请求当前服务外部资源，转发只能请求当前服务内部资源。

（3）重定向地址栏会发生变化，转发地址栏不会发生变化。

（4）重定向是响应对象调用方法，转发是请求对象调用方法。

【案例 5-4】通过重定向方式进行页面跳转

下面的代码通过/redirect 虚拟路径请求 Demo02Servlet，地址栏请求是 get 方式，所以

Demo02Servlet 的 doGet()方法会被执行，doGet()重定向到百度首页。

```java
package servlet;
//省略包导入代码
@WebServlet("/redirect")
public class Demo02Servlet extends HttpServlet {
    protected void doGet(HttpServletRequest req, HttpServletResponse resp)
throws ServletException, IOException {
        resp.sendRedirect("https://www.baidu.com");
    }
}
```

重启服务器后通过浏览器访问 http://localhost:8080/redirect，发现打开的是百度首页，地址栏的地址也变成了 https://www.baidu.com。通过浏览器调试模式查看/redirect 请求，可见响应状态码是 302，表示重定向，location 地址是百度首页。紧跟着/redirect 请求浏览器发送了百度首页的请求，如图 5-16 所示。

图 5-16　重定向应用示例

【案例 5-5】通过转发的方式进行页面跳转

下面的代码通过/forward 虚拟路径请求 Demo03Servlet，地址栏请求是 get 方式，所以 Demo03Servlet 的 doGet()方法会被执行，doGet()方法转发到当前服务根目录的 index.jsp。

```java
package servlet;
//省略包导入代码
@WebServlet("/forward")
public class Demo03Servlet extends HttpServlet {
    protected void doGet(HttpServletRequest req, HttpServletResponse resp)
throws ServletException, IOException {
        RequestDispatcher rd=req.getRequestDispatcher("/index.jsp");
        rd.forward(req,resp);
    }
}
```

重启服务器后通过浏览器访问 http://localhost:8080/forward，发现打开的是 index.jsp 的内容，但是地址栏没有变化。通过浏览器调试模式查看请求数据，浏览器只发送了一次 forward 请求，状态码返回 200，表示请求正常响应，如图 5-17 所示。

图 5-17 转发应用示例

5.1.8 作用域与数据共享

Web 应用程序开发过程中经常需要在不同的资源间共享数据，Servlet 提供了 4 种不同范围的作用域，可以让 Servlet/JSP 相互共享数据。4 种作用域及其作用范围如表 5-3 所示。

表 5-3 作用域对象

作用域	类型	作用范围
pageContext	javax.servlet.jsp.JspContext	只在当前 JSP 页面内可以访问
request	javax.servlet.ServletRequest 及子类	只在当前请求有效期间可以访问
session	javax.servlet.http.HttpSession	只在当前会话有效期间可以访问
application	javax.servlet.ServletContext	只在当前 Web 应用程序运行时可以访问

HTTP 本身是无状态的，它的每个请求都是完全独立的，每个请求包含了处理这个请求所需的完整的数据，发送请求不涉及状态变更。绝大部分 Web 应用程序需要将客户端一系列的请求相互关联，例如网上商城应用程序可跨请求保存用户购物车的状态，基于 Web 的应用程序使用一种称为"会话（session）"的方式维护这种状态。为了支持需要维护状态的应用程序，Servlet 技术提供了用于管理会话的 API。

Servlet 里面的会话由 javax.servlet.http.HttpSession 对象表示，会话通过请求对象的 getSession() 方法获取。不同客户端的会话对象是不一样的，会话对象在失效前会一直关联同一个客户端的请求。

会话对象存在于服务器端内存，如果客户端超过设定时间没有请求服务器资源或者会话对象调用了 invalidate() 方法，则可以销毁会话对象。

ServletContext 也称为 Servlet 上下文，每一个 Web 应用程序启动时服务器都会为它创建一个 ServletContext 对象。ServletContext 对象是全局共享的，即不同客户端访问的是同一个 ServletContext 对象，同一个 Web 应用程序下不同 Servlet 访问的是同一个 ServletContext 对象。

所有作用域对象都能通过以下方法进行数据共享访问。

void setAttribute(String k ,Object v)：将数据 v 通过关键字 k 放入作用域对象。

Object getAttribute(String k)：通过关键字 k 从作用域中获取数据值并返回。

【案例 5-6】不同作用域对象的使用（show_scope.jsp）

获取作用域中的 msg 值，并通过<%= %>标签将值输出到页面。

```jsp
<%@ page contentType="text/html;charset=UTF-8" language="java" %>
<html>
<head>
    <title>show_scope.jsp</title>
</head>
<body>
<a href="/scope">请求/scope</a><br>
request 作用域:<%= request.getAttribute("msg")%><br>
session 作用域:<%= session.getAttribute("msg")%><br>
application 作用域:<%=application.getAttribute("msg")%>
</body>
</html>
```

Demo04Servlet 向多个作用域对象添加属性 msg，然后通过转发的方式返回/show_scope.jsp 页面，代码如下所示。

```java
package servlet;
//省略包导入代码
@WebServlet("/scope")
public class Demo04Servlet extends HttpServlet {
    protected void doGet(HttpServletRequest req, HttpServletResponse resp) throws ServletException, IOException {
        req.setAttribute("msg","request scope");
        HttpSession session=req.getSession(true);
        session.setAttribute("msg","session scope");
        ServletContext application=req.getServletContext();
        application.setAttribute("msg","application scope");
        req.getRequestDispatcher("/show_scope.jsp").forward(req,resp);
    }
}
```

重启服务器后通过浏览器访问 http://localhost:8080/show_scope.jsp，可以看见作用域输出内容都是 null，这是因为这时还没有为作用域 msg 属性赋值。

单击超链接"请求/scope"，执行 Demo01Servlet 的 doGet()方法为作用域赋值并转发，回到/show_scope.jsp 页面，所有作用域都输出各自的 msg 属性值，如图 5-18 所示。

使用当前浏览器将地址栏地址改为 http://localhost:8080/show_scope.jsp 重写请求，页面中 request 作用域的 msg 属性值变为 null，session 和 application 作用域的 msg 属性值还是正常输出，如图 5-19 所示。这是因为重新请求/show_scope.jsp 会创建一个新的请求对象，这个新请求作用域对象没有设置 msg 属性，而 session 因为是同一个客户端的请求，所以保持不变；application 是上下文，也保持不变。

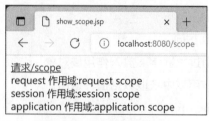

图 5-18 作用域的使用（1）

图 5-19 作用域的使用（2）

重新打开一个浏览器程序来访问 http://localhost:8080/show_scope.jsp 页面，这时只有 application 作用域的 msg 属性值有内容。这是因为重新打开的浏览器窗口是一个新的客户端，它对应的 session 对象还没有设置 msg 属性。application 是全局共享的，它的 msg 属性在前面的操作中设置了值。

5.2 JSP 基础

5.2.1 JSP 运行原理

JavaServer Pages 的缩写为 JSP，它是对 Servlet 技术的扩展。一个 JSP 页面含有 HTML 标签和 JSP 标签，使用 JSP 可以更方便地创建动态页面。

JSP 具有以下优点。

（1）扩展 Servlet。JSP 是 Servlet 技术的扩展，因此可以使用 Servlet 所有的特性。

（2）代码分层。JSP 可以让程序员很方便地将业务代码和展示代码分离。

（3）减少代码。JSP 中可以使用标签、EL 表达式，以及内置对象等来减少代码量。

JSP 是对 Servlet 的扩展，它的完整运行过程如下。

（1）将 JSP 文件转换成 Servlet 文件。

转换过程中 HTML 标签会转换成 Servlet 中的输出语句，JSP 标签会根据各自的语法转换成不同的 Java 代码，最后将内容合并成一个完整的 Servlet 文件。

（2）编译转换的 Servlet 文件。

编译过程会根据 Java 语法检查生成的 Servlet 文件，如果存在语法错误，则编译失败，服务器会返回错误码 500，编译通过后得到可执行的字节码。

（3）加载并初始化。

编译通过后，Servlet 容器会加载编译后的字节码来创建 Servlet 对象，执行其中的 jspInit() 方法完成初始化。

（4）处理请求并响应。

Servlet 容器将请求和响应对象作为参数来调用 Servlet 实例的_jspService()方法。该方法会根据转换的代码执行，并将结果通过输出流输出到客户端浏览器。

（5）销毁资源。

如果容器需要移除 JSP 页面的 Servlet 实例，则会调用它的 jspDestroy()方法。

JSP 页面的 Servlet 实例创建完成后会保留在容器中，当再次请求该 JSP 页面时会直接执行第（4）步。如果该 JSP 源代码发生变化，再次请求时会按照完整的流程重新执行。

5.2.2 JSP 内置对象

在 JSP 中可以使用 Servlet API，其中一些常用 API 在转换过程中已经声明并赋值完成，可在 JSP 中直接使用，这些 API 对象称为 JSP 的内置对象，如表 5-4 所示。

表 5-4 JSP 内置对象

对象	描述
request	javax.servlet.http.HttpServletRequest 接口的实例
response	javax.servlet.http.HttpServletResponse 接口的实例
out	javax.servlet.jsp.JspWriter 类的实例，用于把结果输出至网页上
session	javax.servlet.http.HttpSession 接口的实例
application	javax.servlet.ServletContext 接口的实例
config	javax.servlet.ServletConfig 接口的实例
pageContext	javax.servlet.jsp.PageContext 类的实例，提供对 JSP 页面所有对象以及命名空间的访问
page	当前 JSP 对象
exception	代表发生错误的 JSP 页面中的异常对象

【案例 5-7】JSP 内置对象的使用

demo01.jsp 中通过<%= %>标签直接输出内置对象并调用相关方法的返回值。JSP 中内置对象可以直接使用。

```jsp
<%@ page contentType="text/html;charset=UTF-8" language="java" %>
<html>
<head>
    <title>demo01.jsp</title>
</head>
<body>
url : <%= request.getRequestURL()%><br>
locale:<%= response.getLocale()%><br>
sessionId : <%=session.getId()%><br>
server info : <%=application.getServerInfo()%><br>
</body>
</html>
```

运行结果如图 5-20 所示。

图 5-20 JSP 内置对象的使用示例

5.2.3 JSP 标签

JSP 提供了多种标签来完成不同的操作，如指令标签、程序片段标签、动作标签等，如表 5-5 所示。

表 5-5 JSP 标签

标签	描述
<% %>	程序片段标签，里面可以放 Java 代码
<%= %>	输出标签，里面可以放表达式或者值
<%! %>	声明标签，里面可以声明全局变量、方法或者类
<%@ page %>	page 指令标签，设置当前 JSP 相关属性
<%@ include %>	include 指令标签，静态导入内容，将内容和当前 JSP 代码合并
<%@ taglib %>	taglib 指令标签，导入标签库
<jsp:xx />	各种动作指令标签，如 forward 转发、include 动态导入等

【案例 5-8】JSP 标签的使用

demo02.jsp 通过 include 指令标签将 movie_list.inc 文件的源代码导入 demo02.jsp 中一起编译运行，代码如下所示。

```jsp
<%@ page contentType="text/html;charset=UTF-8" language="java" %>
<%@ page import="java.util.List,java.util.ArrayList" %>
<html>
<head>
    <title>demo02.jsp</title>
</head>
<body>
<%
    List<String> list=new ArrayList<String>();
    list.add("董存瑞");
    list.add("上甘岭");
    list.add("冰山上的来客");
    list.add("英雄儿女");
    list.add("开国大典");
%>
<%@include file="movie_list.inc" %>
</body>
</html>
```

movie_list.inc 文件是被导入的文件，代码如下所示。

```jsp
<%@ page contentType="text/html;charset=UTF-8" language="java" %>
红色电影列表:<hr/>
<ul>
<%
    for (String movie:list) {
%>
    <li><%= movie %></li>
<%
    }
%>
</ul>
```

在项目开发中可以使用 include 指令标签将公共部分代码分离成被导入文件，提高代码重用性。

page 指令标签的 contentType 属性用于设置文件的内容和编码方式，如果 include 指令标签导入的文件中含有中文，则需要在导入文件里面也加上 page 指令标签来设置 contentType 属性指定编码方式，以避免乱码。

page 指令标签的 import 属性可以在 JSP 中导入 Java 的类，如果有多个类，可以使用逗号隔开，或者使用多个 page 指令标签的 import 属性。

程序片段标签可以将 Java 代码和 HTML 代码混合，例如代码中循环 list 对象生成多个<li>标签。

运行 demo02.jsp，结果如图 5-21 所示。

图 5-21　JSP 标签的应用示例

5.3　EL 表达式和 JSTL

5.3.1　EL 表达式

Expression Language 的缩写为 EL，它简化了对象属性的访问方式。EL 支持 EL 内置对象、运算符和关键字。

EL 支持的内置对象如表 5-6 所示，和 JSP 的内置对象有所不同。

表 5-6　EL 支持的内置对象

内置对象	说明
pageScope	获取 page 作用域的属性
requestScope	获取 request 作用域的属性
sessionScope	获取 session 作用域的属性
applicationScope	获取 application 作用域的属性
param	获取请求中参数的一个值
paramValues	获取请求中参数的一组值
header	获取请求头中的一个值
headerValues	获取请求头中的一组值
cookie	获取 cookie 对象
initParam	获取初始化参数值
pageContext	JSP 的 pageContext 对象，通过它可以访问原生的作用域对象

EL 表达式的基本语法为${表达式}，其作用是访问表达式的值并返回输出。

EL表达式中可以使用多种运算符和关键字。

（1）访问运算符：.、[]。例如${user.username}、${user['username']}。使用EL表达式访问属性前如果没有给出作用域，则它会依次查找page、request、session和application作用域，如果找到就返回。

（2）算术运算符：+、-、*、/、div、%、mod。EL算术运算可以使用符号或者对应的关键字，例如${5/2}、${5 div 2}。

（3）逻辑运算符：and、&&、or、||、not、!。

（4）关系运算符：==、eq、!=、ne、<、lt、>、gt、<=、ge、>=、le。关系运算符可以用于布尔、数字、文本的比较，例如${(10*10) eq 100}、${'a' < 'b'}。

（5）空运算符：empty。判断值是否为null或者内容是否为空。例如${!empty param.id}，${empty requestScope.userList}。

（6）三目运算符：A?B:C。表达式A成立则返回表达式B，否则返回表达式C。例如${user.role==0?"管理员":"一般用户"}。

【案例5-9】EL表达式的使用

demo03.jsp分别创建两个User对象并将其放到page和session作用域，然后通过EL的内置对象访问相关的值，代码如下所示。

```
<%@ page import="po.User" %>
<%@ page contentType="text/html;charset=UTF-8" language="java" %>
<html>
<head>
    <title>demo03.jsp</title>
</head>
<body>
<%
    User user1=new User();
    user1.setId(1);
    user1.setUsername("张三");
    user1.setRole(0);
    pageContext.setAttribute("user",user1);
    User user2=new User();
    user2.setId(2);
    user2.setUsername("李四");
    user2.setRole(1);
    session.setAttribute("user",user2);
%>
pageScope.user : <br>
id : ${pageScope.user.id} , name : ${pageScope.user.username} ,role : ${pageScope.user.role==0?"管理员":"普通用户"}<br>
sessionScope.user : <br>
id : ${sessionScope.user.id} , name : ${sessionScope.user.username} ,role : ${sessionScope.user.role==0?"管理员":"普通用户"}<br>
applicationScope.user ${empty applicationScope.user?"不存在":"存在"}<br>
User-Agent : <br>
${header["User-Agent"]}<br>
</body>
</html>
```

User.java 是用户数据的封装类，代码如下所示。

```java
package po;
public class User {
    private int id;
    private String username;
    private int role;
    public int getId() {
        return id;
    }
    public void setId(int id) {
        this.id = id;
    }
    public String getUsername() {
        return username;
    }
    public void setUsername(String username) {
        this.username = username;
    }
    public int getRole() {
        return role;
    }
    public void setRole(int role) {
        this.role = role;
    }
}
```

代码运行结果如图 5-22 所示。

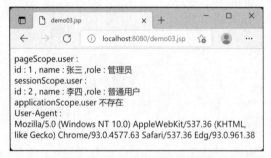

图 5-22　EL 表达式应用示例

5.3.2　JSTL core 标签库

JavaServer Pages Standard Tag Library 的缩写为 JSTL，它封装了许多 JSP 应用程序所共用的核心功能。JSTL 由 Apache 的 Jakarta 小组来维护，主要为 Java Web 开发人员提供一个标准通用的标签库。JSTL 能够简化 JSP 开发过程，避免使用容易出错的程序片段标签。

JSTL 有 5 种标签，分别是 core 标签、function 标签、formatting 标签、xml 标签和 sql 标签。这里仅要求掌握 core 标签。

使用 JSTL 需要下载相关的库文件，访问 Tomcat 官网的 JSTL 下载页面，下载 Impl 和 Spec 两个 .jar 包，如图 5-23 所示。

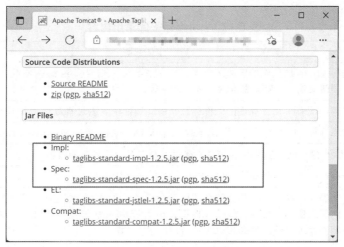

图 5-23 JSTL 下载页面

下载完成后在项目的 web/WEB-INF 文件夹下新建 lib 文件夹,将下载完成的.jar 文件复制进去。选中两个.jar 文件并单击鼠标右键,在弹出的快捷菜单中选择"Add as Library…"命令,如图 5-24 所示。

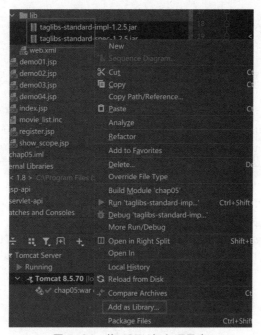

图 5-24 将 JSTL 加入项目库

想在 JSP 中使用 JSTL,还需要在 JSP 页面中通过 taglib 指令标签引入标记库。下面的代码用于引入 JSTL 的 core 标签库,uri 的值是通过标记库的 tld 文档定义的。代码设置前缀 c,这样当前 JSP 页面就可以通过<c:xx >标签来使用 core 标签库。

```
<%@ taglib prefix="c" uri="http://java.sun.com/jsp/jstl/core" %>
```

core 标签库常用内容如表 5-7 所示。

表 5-7　core 标签库常用内容

标签	说明
c:out	输出 value 属性中的表达式结果，与 <%= %> 相似。escapeXML 属性设置标签是否需要转义输出，默认是 true，表示转义输出
c:if	含有测试条件 test 属性的条件标签，test 属性中表达式值为 true 时显示内容
c:choose	选择标签，choose 标签是父标签，用于定义一组选择，when 和 otherwise 是子标签
c:when	when 含有测试条件 test 属性，条件为真时，执行对应的内容
c:otherwise	所有 when 条件都不成立时，执行 otherwise 的内容
c:forEach	含有循环对象 items 属性的循环标签，每次从 items 中取出一个元素赋值给 var 属性指定的临时变量，在循环标签内部可以访问临时变量，通过 varStatus 属性指定的变量可以访问循环状态，如索引 index、计数 count 等
c:param	为 c:url 标签生成的 URL 添加参数名值对
c:url	创建一个 URL

【案例 5-10】 core 标签库的使用

demo04.jsp 中 JSP 页面根据请求作用域中是否存在查询结果 userList 来呈现不同的页面，代码如下所示。

```jsp
<%@ page contentType="text/html;charset=UTF-8" language="java" %>
<%@ taglib prefix="c" uri="http://java.sun.com/jsp/jstl/core" %>
<html>
<head>
    <title>demo04.jsp</title>
</head>
<body>
<c:if test="${empty requestScope.userList}">
    <a href="<c:url value="/users"></c:url>">查询用户列表</a>
</c:if>

<c:choose>
    <c:when test="${not empty requestScope.userList}">
        <ul>
            <c:forEach var="user" items="${requestScope.userList}" varStatus="vs">
                <li><c:out value="${vs.count}"/>. <c:out value="${user.username}"/> - <c:out value="${user.role==0?'管理员':'普通用户'}"/></li>
            </c:forEach>
        </ul>
    </c:when>
    <c:otherwise>
        <h5>用户列表为空，没有数据</h5>
    </c:otherwise>
</c:choose>
</body>
</html>
```

Demo05Servlet.java 是一个 Servlet，用于生成一个 User 集合对象并将其放入 request 作用域，通过转发的方式跳转回/demo04.jsp，代码如下所示。

```java
package servlet;

import po.User;

import javax.servlet.ServletException;
import javax.servlet.annotation.WebServlet;
import javax.servlet.http.HttpServlet;
import javax.servlet.http.HttpServletRequest;
import javax.servlet.http.HttpServletResponse;
import java.io.IOException;
import java.util.ArrayList;
import java.util.List;

@WebServlet("/users")
public class Demo05Servlet extends HttpServlet {
    protected void doGet(HttpServletRequest req, HttpServletResponse resp) throws ServletException, IOException {
        List<User> list = new ArrayList<User>();
        User user1=new User();
        user1.setId(1);
        user1.setUsername("张三");
        user1.setRole(0);
        list.add(user1);

        User user2=new User();
        user2.setId(2);
        user2.setUsername("李四");
        user2.setRole(1);
        list.add(user2);

        User user3=new User();
        user3.setId(3);
        user3.setUsername("王五");
        user3.setRole(1);
        list.add(user3);

        req.setAttribute("userList",list);
        req.getRequestDispatcher("/demo04.jsp").forward(req,resp);
    }
}
```

core 标签库的 test 属性和 value 属性都支持通过 EL 表达式赋值。

运行 demo04.jsp，由于没有数据，会显示查询链接和提示，单击链接调用 Servlet，通过 request 作用域传递数据并返回 demo04.jsp，显示结果发生变化。代码运行结果如图 5-25 所示。

图 5-25 core 标签库的应用示例

5.4 过滤器

5.4.1 过滤器简介

过滤器(Filter)可以拦截请求,在请求到达资源之前和请求完成后运行。使用过滤器后 Servlet 可以更专注于业务操作,将一些非业务操作交给过滤器。过滤器主要用于执行转换、记录日志、加密和解密、输入验证等过滤任务。

过滤器在使用时需要将过滤器类和资源地址进行映射,映射完毕后请求资源会首先执行匹配映射的过滤器,在过滤器中可以根据需要让请求继续往下执行或者跳转到其他资源。

同一个资源地址可以匹配多个过滤器映射,过滤器会按照配置顺序(如果是注解配置,则按照类名顺序)依次执行,直到资源请求结束开始响应,响应过程也会经过过滤器,如图 5-26 所示。

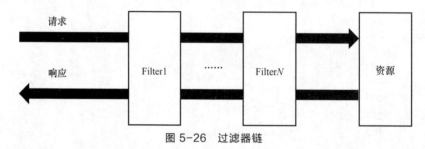

图 5-26 过滤器链

与过滤器相关的 API 有 Filter、FilterChain 和 FilterConfig 3 个,都位于 javax.servlet 包下。通过实现 Filter 接口可以自定义一个过滤器。

Filter 接口中有 init()、destroy()和 doFilter()3 个方法需要实现。其中 init()方法在过滤器初始化时被容器执行,destroy()方法在过滤器需要移除时给容器执行。最重要的是 doFilter()方法,每次映射的资源被请求都会执行相关过滤器的 doFilter()方法。

```
public void doFilter(ServletRequest req, ServletResponse resp, FilterChain
filterChain)
        throws IOException, ServletException { }
```

doFilter()方法的前两个参数分别是 ServletRequest 和 ServletResponse,它们不包含 HTTP 相关方法。如果需要执行 HTTP 相关方法,需要将它们强制转换成 HttpServletRequest 和 HttpServletResponse 对象再使用。

doFilter()方法的第 3 个参数 FilterChain 接口对象也具有 doFilter()方法,能够让请求按照过滤器链继续往下执行。

过滤器映射可以分为路径映射和类型映射。

路径映射以/符号开始,/符号代表当前 Web 应用程序根目录,后面跟虚拟路径,例如 /login、/servlet/hello、/*。

类型映射以.文件类型后缀结尾，例如.jsp，.html。

映射支持通配符*，它代表任意字符。

5.4.2 过滤器使用

【案例 5-11】使用过滤器处理中文乱码

EncodingFilter.java 是一个过滤器，在 doFilter()方法中对请求对象和响应对象设置编码，实现中文乱码的处理，代码如下所示。

```java
package filter;
//省略包导入代码
@WebFilter("/*")
public class EncodingFilter implements Filter {
    public void init(FilterConfig filterConfig) throws ServletException {
    }
    public void doFilter(ServletRequest req, ServletResponse resp, FilterChain filterChain) throws IOException, ServletException {
        req.setCharacterEncoding("utf-8");
        resp.setContentType("text/html;charset=utf-8");
        filterChain.doFilter(req,resp);
    }
    public void destroy() {
    }
}
```

将 Demo01Servlet 的 doPost()方法里处理中文乱码的代码注释掉或者删除，重新启动服务器运行 http://localhost:8080/register.jsp，提交表单后查看结果。虽然 Demo01Servlet 中没有了处理乱码的代码，但是中文依旧可以正常显示。这是因为/*映射会过滤所有的请求，包括 Demo01Servlet 的/demo01。进入 EncodingFilter 的 doFilter()方法后会对请求和响应对象进行乱码处理，然后将处理过的请求和响应对象通过过滤器链传递，所以最后交给 Demo01Servlet 的请求和响应对象是已经进行过中文乱码处理的。

5.5 MVC 模式

MVC 是 Model View Controller 的缩写。MVC 是一种设计模式，用来分离业务代码、展示代码和数据。

Model 是模型层，可以进一步细分为业务模型和数据模型。业务模型是指实现业务逻辑的代码，数据模型用于封装系统的数据状态。

View 是视图层，指用来展示效果的代码，一般是用户交互 UI 代码，例如 JSP。

Controller 是控制层，用来连接 Model 和 View。控制层代码接收视图层的请求，获取传递的数据并调用相关的模型层，最后将模型层的结果返回给对应的视图层。

在 MySQL 中新建数据库 chap05，在 chap05 下新建分类表 t_category，表结构如表 5-8 所示。

表 5-8　t_category 表结构

字段	类型	约束	允许为空	其他	说明
id	int	PK	NO	auto_increment	分类主键
name	varchar(45)		NO		分类名
description	text		YES		分类描述

【案例 5-12】按照 MVC 模式完成 t_category 表数据的浏览和添加

第一步完成数据模型 Catagory.java，模型层用来封装数据，代码如下所示。

```java
package po;
public class Category {
    private int id;
    private String name;
    private String description;
    public int getId() {
        return id;
    }
    public void setId(int id) {
        this.id = id;
    }
    public String getName() {
        return name;
    }
    public void setName(String name) {
        this.name = name;
    }
    public String getDescription() {
        return description;
    }
    public void setDescription(String description) {
        this.description = description;
    }
}
```

第二步完成数据访问层 CategoryDAO.java，数据访问层用来和数据库交互。getAll()方法用于查询 t_category 表，并将查询结果封装成 Category 对象以集合的方式返回；save(Category c)方法用于将 Category 对象保存到 t_category 表中。编写代码前参考添加 JSTL 库文件的方式，将 MySQL 驱动放到/WEB-INF/lib 文件夹下，然后添加到项目运行库中，代码如下所示。

```java
package dao;
//省略包导入代码
public class CategoryDAO {
    public static final String JDBC_URL = "jdbc:mysql://localhost:3306/chap05?useUnicode=true&characterEncoding=utf-8&serverTimezone=UTC&allowMultiQueries=true";
    public static final String JDBC_USER = "root";
    public static final String JDBC_PASSWORD = "123456";

    public List<Category> getAll() throws Exception {
        List<Category> list = new ArrayList<Category>();
        Connection con = null;
        PreparedStatement ps = null;
```

```java
            ResultSet rs = null;
            try {
                Class.forName("com.mysql.cj.jdbc.Driver");
                con = DriverManager.getConnection(JDBC_URL, JDBC_USER, JDBC_PASSWORD);
                ps = con.prepareStatement("select  * from t_category");
                rs = ps.executeQuery();
                while (rs.next()) {
                    Category c = new Category();
                    c.setId(rs.getInt("id"));
                    c.setName(rs.getString("name"));
                    c.setDescription(rs.getString("description"));
                    list.add(c);
                }
            } catch (Exception e) {
                throw new Exception("DAO异常:" + e.getMessage());
            } finally {
                if (rs != null) {
                    rs.close();
                }
                if (ps != null) {
                    ps.close();
                }
                if (con != null) {
                    con.close();
                }
            }
            return list;
        }


        public void save(Category c) throws Exception {
            Connection con = null;
            PreparedStatement ps = null;
            try {
                Class.forName("com.mysql.cj.jdbc.Driver");
                con = DriverManager.getConnection(JDBC_URL,JDBC_USER,JDBC_PASSWORD);
                ps = con.prepareStatement("insert into t_category value(null,?,?)");
                ps.setString(1, c.getName());
                ps.setString(2, c.getDescription());
                ps.executeUpdate();
            } catch (Exception e) {
                throw new Exception("DAO异常:" + e.getMessage());
            } finally {
                if (ps != null) {
                    ps.close();
                }
                if (con != null) {
                    con.close();
                }
            }
        }
}
```

第三步分别完成两个功能的控制层代码。

浏览的控制层代码是 ListCategory，处理对/category/list 的 get 请求，通过 request 作用域返回数据模型 Category 的集合给视图层，代码如下所示。

```java
package servlet;
//省略包导入代码
@WebServlet("/category/list")
public class ListCategory extends HttpServlet {
    protected void doGet(HttpServletRequest req, HttpServletResponse resp) throws ServletException, IOException {
        try {
            List<Category> list=new CategoryDAO().getAll();
            req.setAttribute("categoryList",list);
            req.getRequestDispatcher("/category.jsp").forward(req,resp);
        } catch (Exception e) {
            e.printStackTrace();
        }
    }
    protected void doPost(HttpServletRequest req, HttpServletResponse resp) throws ServletException, IOException {
        doGet(req, resp);
    }
}
```

添加的控制层代码是 AddCategoryServlet，处理对/category 的 post 请求。将请求参数值封装成数据模型 Category，交给 DAO 保存。添加完成后为了显示最新的数据，转发/category/list 查询最新数据并返回视图层，代码如下所示。

```java
package servlet;
//省略包导入代码
@WebServlet("/category")
public class AddCategory extends HttpServlet {
    protected void doPost(HttpServletRequest req, HttpServletResponse resp) throws ServletException, IOException {
        try {
            Category c = new Category();
            c.setName(req.getParameter("name"));
            c.setDescription(req.getParameter("description"));
            new CategoryDAO().save(c);
            req.getRequestDispatcher("/category/list").forward(req,resp);
        } catch (Exception e) {
            e.printStackTrace();
        }
    }
}
```

第四步完成视图层 category.jsp。视图层中加入了 Bootstrap 效果，需要根据前面的内容添加相关 CSS 文件到当前项目中，代码如下所示。

```jsp
<%@ page contentType="text/html;charset=UTF-8" language="java" %>
<%@ taglib prefix="c" uri="http://java.sun.com/jsp/jstl/core" %>
<html>
<head>
```

```html
        <title>分类管理</title>
        <link rel="stylesheet" href="/css/bootstrap.min.css">
</head>
<body>
<div class="container mt-5">
    <div class="row">
        <div class="col-4">
            <form action="/category" method="post">
                <div class="form-group">
                    <label>分类名</label>
                    <input type="text" class="form-control" name="name">
                </div>
                <div class="form-group">
                    <label>分类描述</label>
                    <textarea class="form-control" name="description"></textarea>
                </div>
                <button type="submit" class="btn btn-primary">添加</button>
                <a class="btn btn-outline-success" href="/category/list">查看所有分类</a>
            </form>
        </div>
        <div class="col">
            <c:if test="${not empty requestScope.categoryList}">
                <table class="table table-striped">
                    <thead>
                    <tr class="">
                        <th scope="col">主键</th>
                        <th scope="col">分类名</th>
                        <th scope="col">分类描述</th>
                    </tr>
                    </thead>
                    <tbody>
                    <c:forEach var="c" items="${requestScope.categoryList}">
                        <tr>
                            <td>${c.id}</td>
                            <td>${c.name}</td>
                            <td>${c.description}</td>
                        </tr>
                    </c:forEach>
                    </tbody>
                </table>
            </c:if>
        </div>
    </div>
</div>
</body>
</html>
```

运行 http://localhost:8080/category.jsp，添加数据或者单击"查看所有分类"按钮，运行结果如图 5-27 所示。

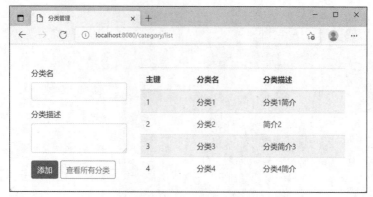

图 5-27　MVC 应用示例

5.6　Servlet 返回 JSON 数据

在异步交互模式下，前端需要的不是一整个页面的内容，而是局部的数据。因此 Servlet 不会以跳转页面的方式进行响应，而是通过输出流对象将数据以文本的方式进行返回，其中最常见的文本返回格式是 JSON。

使用 Servlet 输出 JSON 格式的数据是一件费时费力的事情，而且容易出错。第三方提供的多种 Java 工具库可以实现 Java 对象和 JSON 数据的转换，例如 GSON、JackSON 和 FastJSON，其中 FastJSON 是阿里巴巴提供的开源工具库。下载 FastJSON 工具库文件，将下载的 fastjson-1.2.76.jar 放到项目的/WEB-INF/lib 文件夹下，并参考 5.3.2 小节的操作将其添加到项目的库文件中。

【案例 5-13】使用 Servlet 接收 AJAX 请求并返回 JSON 数据

category_json.jsp 使用 jQuery 绑定链接的单击事件异步请求/json/category/list 对应的 Servlet 查询数据。由于是异步请求，Servlet 不再跳转页面，而是以 JSON 格式返回数据。所以需要设置 ContentType 为 application/json，这样 jQuery 拿到的返回值就可以直接当作 JSON 对象使用，不用再转换，代码如下所示。

```jsp
<%@ page contentType="text/html;charset=UTF-8" language="java" %>
<html>
<head>
    <title>分类管理</title>
    <link rel="stylesheet" href="/css/bootstrap.min.css">
</head>
<body>
<div class="container mt-5">
    <div class="row">
        <div class="col-4">
            <a id="load_category" class="btn btn-outline-success">查看所有分类</a>
        </div>
        <div class="col">

            <table class="table table-striped">
```

```html
            <thead>
                <tr class="">
                    <th scope="col">主键</th>
                    <th scope="col">分类名</th>
                    <th scope="col">分类描述</th>
                </tr>
            </thead>
            <tbody>
            </tbody>
        </table>

    </div>
  </div>
</div>
<script src="js/jquery-3.6.0.min.js"></script>
<script>
    $(function () {
        $("#load_category").click(function () {
            $.getJSON("/json/category/list", function (result) {
                let category_list = result.data;
                $("tbody").empty();
                category_list.forEach(function ({id, name, description}) {
                    $("tbody").append('<tr><td>\${id}</td><td>\${name}</td><td>\${description}</td></tr>');
                })
            });
        });
    });
</script>
</body>
</html>
```

FastJSON 提供了一个 JSONObject 类，通过 put()方法可以将需要转换的 Java 对象加进去。JSONObject 对象准备完毕后通过 JSON.toJSONString()方法将它的内容转换为 JSON 格式的字符串。Demo06Servlet 将转换后的 JSON 数据通过输出流输出，代码如下所示。

```java
package servlet;
//省略包导入代码
@WebServlet("/json/category/list")
public class Demo06Servlet extends HttpServlet {
    protected void doGet(HttpServletRequest req, HttpServletResponse resp) throws ServletException, IOException {
        resp.setContentType("application/json");
        PrintWriter out=resp.getWriter();
        try {
            List<Category> list=new CategoryDAO().getAll();
            JSONObject jsonObj=new JSONObject();
            jsonObj.put("data",list);
            jsonObj.put("msg","操作成功");
            String jsonStr= JSON.toJSONString(jsonObj);
            out.write(jsonStr);
            out.flush();
            out.close();
```

```
            } catch (Exception e) {
                e.printStackTrace();
            }
        }
    }
```

通过浏览器的调试工具可以看到 Servlet 返回 JSON 格式的结果，如图 5-28 所示。

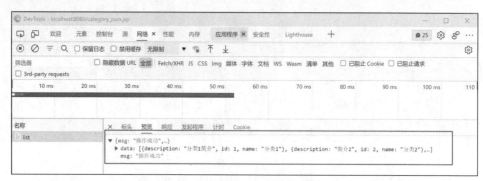

图 5-28　返回 JSON 数据

运行 http://localhost:8080/category_json.jsp，单击"查看所有分类"按钮异步加载数据，再通过 jQuery 动态生成表格内容，结果如图 5-29 所示。

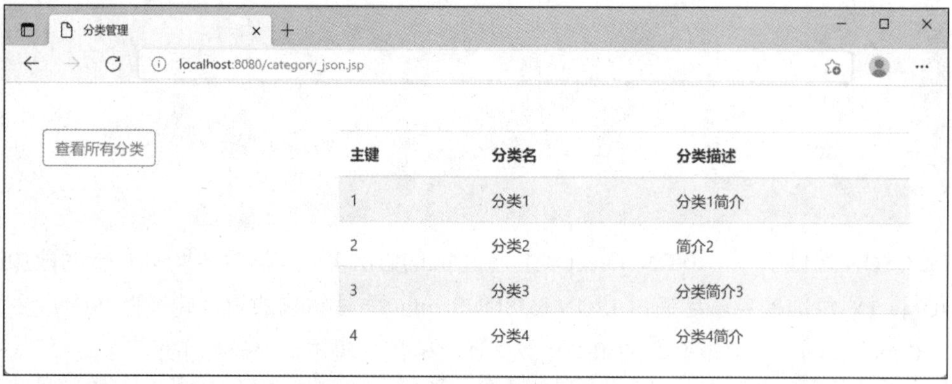

图 5-29　异步加载动态数据

本章习题

1. 请简述 Servlet 运行原理。
2. 请简述 JSP 运行原理。
3. 新建 Web 项目并添加中文乱码过滤器。
4. 在习题 3 的基础上，结合第 4 章习题的 po 类和 dao 类，新建 JSP 页面和 Servlet 类，实现基于 MVC 模式的 t_category 表的添加、删除、修改和查询操作。

实战篇

第6章
项目需求与设计

本章目标

- 理解超市管理系统需求
- 理解超市管理系统设计
- 完成项目启动工作

6.1 项目概述

超市是生活中常见的商业形态,以往的小型超市信息管理依赖于人工,工作量大且容易出错。本章围绕超市中常见的需求,如商品分类信息、供应商信息、商品信息、进货信息和销售信息管理,进行超市管理系统的分析与设计。

超市管理系统需要具备以下功能,如表 6-1 所示。

表 6-1 超市管理系统功能需求

用例编号	用例功能	说明
01	查询供应商	管理员输入条件查询供应商
02	添加供应商	管理员通过供应商添加页面添加新的供应商(供应商名、联系人、联系电话、简介)
03	删除供应商	管理员删除指定供应商
04	修改供应商	管理员对供应商信息进行编辑后将数据更新到数据库
05	查询分类	管理员输入条件查询商品分类
06	添加分类	管理员通过分类添加页面添加新的商品分类(分类名、分类介绍)
07	删除分类	管理员删除指定分类
08	修改分类	管理员修改指定分类
09	查询商品	管理员输入条件查询商品
10	添加商品	管理员通过商品添加页面添加新的商品(商品名、规格、条形码、售价、分类)
11	删除商品	管理员删除指定商品
12	修改商品	管理员修改指定商品
13	添加进货记录	管理员添加一条进货记录(商品、供应商、进货日期、进货价格、进货数量)
14	查询进货记录	管理员输入条件查询进货记录
15	添加销售记录	管理员添加一条销售记录(销售时间、商品、数量)
16	查询销售记录	管理员输入条件查询销售记录
17	查看销售明细	管理员查看销售记录对应的销售明细

上述功能均需要以管理员身份实现。

超市管理系统的功能结构图如图 6-1 所示。

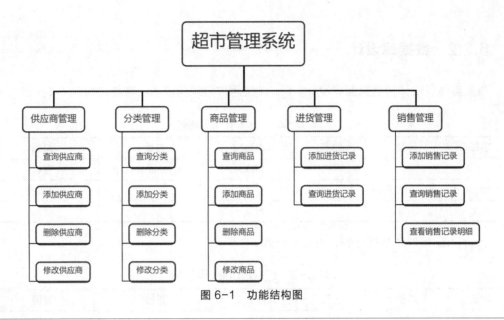

图 6-1 功能结构图

6.2 系统设计

6.2.1 实体类设计

根据超市管理系统需求，抽象出系统实体类，分别是 User（管理员）、Supplier（供应商）、Category（商品分类）、Goods（商品）、Restock（进货记录）、Sale（销售记录）、SaleItem（销售明细），如图 6-2 所示。

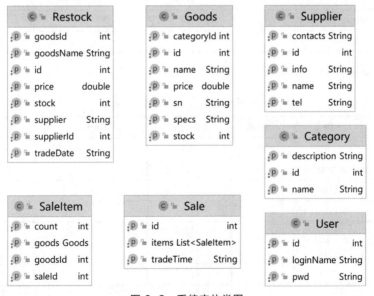

图 6-2 系统实体类图

6.2.2 数据库设计

根据实体类设计得到数据库设计。t_user 是用户登录信息表，表结构如表 6-2 所示。

表 6-2 t_user 表结构

字段	类型	约束	允许为空	其他	说明
id	int	PK	NO	auto_increment	主键
loginName	varchar(45)		NO		登录名
pwd	varchar(45)		NO		密码

t_supplier 是供应商信息表，表结构如表 6-3 所示。

表 6-3 t_supplier 表结构

字段	类型	约束	允许为空	其他	说明
id	int	PK	NO	auto_increment	主键
name	varchar(45)		NO		供应商名
contacts	varchar(45)		NO		联系人
tel	varchar(45)		NO		联系电话
info	text		YES		简介

t_category 是分类信息表，表结构如表 6-4 所示。

表 6-4 t_category 表结构

字段	类型	约束	允许为空	其他	说明
id	int	PK	NO	auto_increment	主键
name	varchar(45)		NO		分类名
description	text		YES		分类描述

t_goods 是商品信息表，表结构如表 6-5 所示。

表 6-5 t_goods 表结构

字段	类型	约束	允许为空	其他	说明
id	int	PK	NO	auto_increment	主键
name	varchar(45)		NO		商品名
specs	varchar(45)		NO		规格
sn	varchar(45)		NO		条形码
price	double		NO		售价
stock	int		NO		库存
category_id	int		NO		分类主键

t_restock 是进货信息表，表结构如表 6-6 所示。

表 6-6　t_restock 表结构

字段	类型	约束	允许为空	其他	说明
id	int	PK	NO	auto_increment	主键
goods_id	int		NO		商品主键
supplier_id	int		NO		供应商主键
price	double		NO		进货价格
stock	int		NO		进货数量
trade_date	date		NO		进货日期

t_sale 是销售信息表，表结构如表 6-7 所示。

表 6-7　t_sale 表结构

字段	类型	约束	允许为空	其他	说明
id	int	PK	NO	auto_increment	主键
trade_time	datetime		NO		销售时间

t_sale_item 是销售明细表，表结构如表 6-8 所示。

表 6-8　t_sale_item 表结构

字段	类型	约束	允许为空	其他	说明
sale_id	int		NO		销售记录主键
goods_id	int		NO		商品主键
count	int		NO		销售数量

6.2.3　页面原型设计

超市管理系统登录页面原型如图 6-3 所示。

图 6-3　登录页面

超市管理系统后台页面原型分为 4 个区域，分别是顶部导航菜单、左侧主菜单、中间内容区和底部信息，如图 6-4 所示。各功能模块页面整体布局和原型保持一致，内容区域根据需要变化。

图 6-4　后台主页面

6.3　项目准备

6.3.1　项目搭建

【任务 6-1】项目搭建

本章开始进入项目开发阶段，前期项目搭建工作分为以下步骤。

（1）建立数据库。

参考 6.2.2 小节内容，在 MySQL 数据库中新建名为 store 的数据库，并根据项目需要新建 7 张数据库表。

（2）新建项目。

参考 5.1.2 小节内容，在 Intellij IDEA 下新建名为 store 的 Web 项目。并将 servlet-api.jar 和 jsp-api.jar 添加到项目库。

（3）添加第三方包。

将 FastJSON、MySQL 驱动和 JSTL 所需的.jar 包放到/WEB-INF/lib 下，并将它们添加到项目库。

（4）添加 Bootstrap 和 jQuery。

在 web 文件夹下新建 statics 文件夹，参考 1.1.4 小节、1.6.1 小节和 3.1.1 小节内容，将相关的 CSS、JS 等文件放到 statics 文件夹对应的子文件夹下。

完成后项目的前端部分 web 文件夹结构如图 6-5 所示。

Java 源码文件夹 src 按照图 6-6 所示的结构新建目录。controller 目录放控制器 Servlet，dao 目录放数据库访问类，filter 目录放乱码过滤器、权限过滤器和分页过滤器，po 目录放数据模型类或者实体类，utils 目录放分页帮助类。

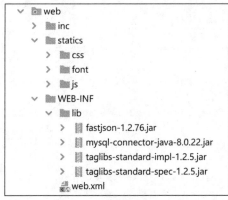

图 6-5　web 文件夹结构图

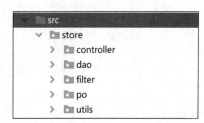

图 6-6　源码文件夹 src 结构图

6.3.2　基础页面实现

【任务 6-2】完成系统主要页面

根据页面原型设计图 6-4，完成后台基础页面编码。多个后台管理页面具有共同的页面布局，其中顶部导航菜单和左侧主菜单在所有页面中是一样的，因此可以将它们分别设计成被导入文件。

在 web/inc 文件夹下新建两个文件 nav.inc 和 menu.inc，nav.inc 放顶部导航菜单代码，menu.inc 放左侧主菜单代码。

顶部导航菜单代码如下所示。

```
<%@ page contentType="text/html;charset=UTF-8" language="java" %>
<nav class="navbar navbar-expand-md bg-dark navbar-dark">
    <a class="navbar-brand" href="/" class="text-muted"><i class="bi-cart"></i>
        超市管理系统</a>
    <button class="navbar-toggler d-lg-none" type="button" data-toggle="collapse" data-target="#collapsibleNavId"></button>
    <div class="collapse navbar-collapse flex-row-reverse" id="collapsibleNavId">
        <ul class="navbar-nav  mt-2 mt-lg-0">
            <li class="nav-item active">
                <a class="nav-link" href="#">欢迎 ${sessionScope.user.loginName}</a>
            </li>
            <li class="nav-item">
                <a class="nav-link" href="/logout"><i class="bi-door-open-fill"></i>退出</a>
            </li>
```

```
            </ul>
        </div>
</nav>
```

左侧主菜单代码如下所示。

```
<%@ page contentType="text/html;charset=UTF-8" language="java" %>
<div class="card">
        <div class="card-header bg-primary text-white">
            功能菜单
        </div>
        <div class="list-group list-group-flush">
            <a href="/supplier/list" class="list-group-item list-group-item-action">供应商管理</a>
            <a href="/category/list" class="list-group-item list-group-item-action">分类管理</a>
            <a href="/goods/list" class="list-group-item list-group-item-action">商品管理</a>
            <a href="/restock/list" class="list-group-item list-group-item-action">进货查询</a>
            <a href="/sale/list" class="list-group-item list-group-item-action">销售管理</a>
        </div>
</div>
```

在后台主页面 index.jsp 中通过静态导入的方式导入 nav.inc 和 menu.inc。index.jsp 代码如下所示。

```
<%@ page contentType="text/html;charset=UTF-8" language="java" %>
<%@taglib prefix="c" uri="http://java.sun.com/jsp/jstl/core" %>
<!DOCTYPE html>
<html lang="en">
<head>
    <meta charset="UTF-8">
    <meta name="viewport" content="width=device-width, initial-scale=1, shrink-to-fit=no">
    <title>管理首页</title>
    <link href="/statics/css/bootstrap.css" rel="stylesheet"/>
    <link href="/statics/font/bootstrap-icons.css" rel="stylesheet">
</head>
<body>
<%@include file="inc/nav.jsp" %>
<div class="container-fluid">
    <div class="row p-3 mt-3">
        <div class="col-3 col-lg-2">
            <%@include file="inc/menu.jsp" %>
        </div>
        <div class="col-9 col-lg-10">
            <div class="jumbotron">
                <h1 class="display-4">超市管理系统</h1>
                <p class="lead">使用 Bootstrap4、jQuery、Servlet、JSTL 实现简单的超市进销存管理系统。</p>
                <hr class="my-4">
                <p>系统混合使用了同步和异步请求。为了提高编辑信息时的操作便利性，编辑
```

前数据采用异步方式获取。以下为异步获取数据的 URI: </p>
```
                <ul>
                    <li>/categories</li>
                    <li>/category</li>
                    <li>/goods</li>
                    <li>/goods/sn</li>
                    <li>/sale/items</li>
                    <li>/supplier</li>
                    <li>/suppliers</li>
                </ul>
            </div>
        </div>
    </div>
</div>
<footer class="fixed-bottom bg-dark text-white">
    <p class="text-center p-3 m-0">
        超市管理系统
    </p>
</footer>
<script src="/statics/js/jquery.js" type="text/javascript"></script>
<script src="/statics/js/bootstrap.js" type="text/javascript"></script>
</body>
</html>
```

6.3.3 实体类实现

【任务 6-3】完成实体类

参考 6.2.1 小节内容,在 store.po 包下完成 7 个实体类定义,下面以 User 类为例。

```
package store.po;
public class User {
    private int id;
    private String loginName;
    private String pwd;
    public int getId() {
        return id;
    }
    public void setId(int id) {
        this.id = id;
    }
    public String getLoginName() {
        return loginName;
    }
    public void setLoginName(String loginName) {
        this.loginName = loginName;
    }
    public String getPwd() {
        return pwd;
    }
    public void setPwd(String pwd) {
        this.pwd = pwd;
    }
}
```

6.3.4 工具类实现

【任务 6-4】完成乱码过滤器 EncodingFilter.java

添加中文乱码过滤器类 store.filter.EncodingFilter，代码如下所示。

```java
package store.filter;
//省略包导入代码
@WebFilter("/*")
public class EncodingFilter implements Filter {
    public void destroy() {
    }
    public void doFilter(ServletRequest req, ServletResponse resp, FilterChain chain) throws ServletException, IOException {
        req.setCharacterEncoding("utf-8");
        resp.setCharacterEncoding("utf-8");
        chain.doFilter(req, resp);
    }
    public void init(FilterConfig config) throws ServletException {
    }
}
```

【任务 6-5】完成分页帮助类 Pager.java 和分页过滤器 PagerFilter.java

store.utils.Pager 封装分页所需要的信息，如总记录数、每页显示多少条记录和当前是第几页，还包括一些用来计算上一页或者下一页是第几页的方法，代码如下所示。

```java
package store.utils;
public class Pager {
    private int total;//总记录数
    private int pageSize = 5;//每页显示记录数
    private int currentPage;//当前页
    public int getTotal() {
        return total;
    }
    public void setTotal(int total) {
        this.total = total;
    }
    public int getPageSize() {
        return pageSize;
    }
    public void setPageSize(int pageSize) {
        this.pageSize = pageSize;
    }
    public int getPageCount() {
        return total % pageSize == 0 ? total / pageSize : total / pageSize + 1;
    }
    public int getCurrentPage() {
        return currentPage;
    }
    public void setCurrentPage(int currentPage) {
```

```
            this.currentPage = currentPage;
    }
    public int getPrevious() {
        return currentPage == 1 ? 1 : currentPage-1;
    }
    public int getNext() {
        return currentPage == getPageCount() ? currentPage : currentPage + 1;
    }
}
```

store.filter.PagerFilter 会对浏览器请求和服务器内部转发进行过滤，对 Pager 对象进行初始化，并将其放到请求域中供 Servlet 或者 JSP 页面使用。通过过滤器初始化 Pager 可以避免 Servlet 或者 JSP 使用分页对象时出现空指针异常，同时也让 Servlet 更专注于分页业务本身。代码如下所示。

```
package store.filter;
//省略包导入代码
@WebFilter(value = "/*",dispatcherTypes ={DispatcherType.REQUEST,DispatcherType.FORWARD })
public class PagerFilter implements Filter {
    public void init(FilterConfig filterConfig) throws ServletException { }
    public void doFilter(ServletRequest servletRequest, ServletResponse servletResponse, FilterChain filterChain) throws IOException, ServletException {
        Pager pager=(Pager) servletRequest.getAttribute("pager");
        if (pager==null) pager=new Pager();
        String currentPage=servletRequest.getParameter("currentPage");
        pager.setCurrentPage(currentPage==null? 1 : Integer.parseInt(currentPage));
        servletRequest.setAttribute("pager",pager);
        filterChain.doFilter(servletRequest,servletResponse);
    }
    public void destroy() { }
}
```

6.3.5 登录和退出实现

登录验证功能时序图如图 6-7 所示。

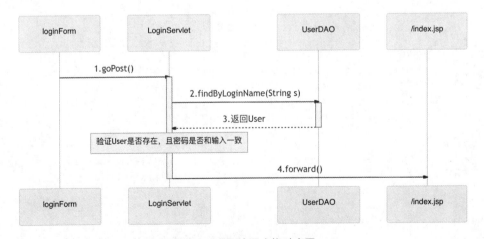

图 6-7 登录验证功能时序图

时序图中各参与者说明如下。

loginForm：视图层，放置于 login.jsp 里面的表单，提交方式为 post，提交地址为 /login，包含控件登录名 loginName 和密码 pwd。

LoginServlet：控制层，位于 store.controller 包中，是一个 Servlet，请求地址为 /login，处理来自 LoginForm 的 post 请求，调用 UserDAO 查询用户信息并验证，如果失败则返回登录页面提示，成功则返回后台首页。

UserDAO：数据访问模型层，其 findByLoginName(String loginName) 方法根据登录名查询 t_user 表，返回对应的 User 对象。

/index.jsp：后台管理主页面的地址。

【任务 6-6】完成数据库连接信息类 DB.java

在 dao 包下新建 DB.java，里面存放数据库连接的相关信息，代码如下所示。

```java
package store.dao;
public class DB {
    public static final String JDBC_DRIVER="com.mysql.cj.jdbc.Driver";
    public static final String JDBC_URL="jdbc:mysql://localhost:3306/store?useUnicode=true&characterEncoding=utf-8&serverTimezone=UTC&allowMultiQueries=true";
    public static final String JDBC_USER="root";
    public static final String JDBC_PASSWORD="123456";
}
```

【任务 6-7】完成方法 UserDAO.findByLoginName(String loginName)

UserDAO 封装对 t_user 表的操作，findByLoginName() 方法通过 loginName 字段查询记录，如果找到就封装 User 对象并返回，没有找到就返回 null，代码如下所示。

```java
package store.dao;
//省略包导入代码
public class UserDAO {
    public User findByLoginName(String loginName) throws Exception {
        User u=null;
        Connection con= null;
        PreparedStatement ps=null;
        ResultSet rs=null;
        try {
            Class.forName(DB.JDBC_DRIVER);
            con= DriverManager.getConnection(DB.JDBC_URL,DB.JDBC_USER,DB.JDBC_PASSWORD);
            ps=con.prepareStatement("select * from t_user where loginname=? ");
            ps.setString(1,loginName);
            rs=ps.executeQuery();
            if (rs.next()){
                u=new User();
                u.setId(rs.getInt("id"));
                u.setLoginName(rs.getString("loginname"));
                u.setPwd(rs.getString("pwd"));
            }
        } catch (Exception e) {
            e.printStackTrace();
            throw new Exception("数据库异常:"+e.getMessage());
```

```
        }finally {
            if(rs!=null) rs.close();
            if(ps!=null) ps.close();
            if(con!=null) con.close();
        }
        return u;
    }
}
```

【任务 6-8】完成 LoginServlet.java

LoginServlet 是控制层,它接收登录表单的 loginName 和 pwd,调用 UserDAO 查询用户信息,并根据返回的 User 对象判断是否登录成功,如果成功,则将信息写入 session。session 中的用户信息可以作为判断登录权限的依据,代码如下所示。

```
package store.controller;
//省略包导入代码
@WebServlet("/login")
public class LoginServlet extends HttpServlet {
    protected void doPost(HttpServletRequest request, HttpServletResponse response) throws ServletException, IOException {
        System.out.println("login");
        String loginName=request.getParameter("loginName");
        String pwd=request.getParameter("pwd");
        try {
            User user=new UserDAO().findByLoginName(loginName);
            if(user==null){
                request.setAttribute("msg","用户名不存在");
                request.getRequestDispatcher("/login.jsp").forward(request,response);
                return;
            }
            if(!user.getPwd().equals(pwd)){
                request.setAttribute("msg","密码错误");
                request.getRequestDispatcher("/login.jsp").forward(request,response);
                return;
            }
            request.getSession(true).setAttribute("user",user);
            request.getRequestDispatcher("/index.jsp").forward(request,response);
        } catch (Exception e) {
            e.printStackTrace();
            throw new RuntimeException(e.getMessage());
        }
    }
}
```

【任务 6-9】完成登录页面 login.jsp

根据页面原型图 6-3 实现系统登录页面 login.jsp,代码如下所示。

```
<%@ page contentType="text/html;charset=UTF-8" language="java" %>
<%@taglib prefix="c" uri="http://java.sun.com/jsp/jstl/core" %>
<!DOCTYPE html>
<html lang="en">
<head>
```

```html
        <meta charset="UTF-8">
        <meta name="viewport" content="width=device-width, initial-scale=1, shrink-to-fit=no">
        <title>登录</title>
        <link href="/statics/css/bootstrap.css" rel="stylesheet"/>
    </head>
    <body class="d-flex justify-content-center align-items-center" style="min-height:600px">

        <div class="bg-light w-50 p-5 shadow d-flex justify-content-center">
            <form id="loginForm" action="/login" method="post" class="w-75">
                <h1>超市管理系统</h1>

                <div class="form-group">
                    <label>账号</label>
                    <input type="text" name="loginName" class="form-control">
                </div>
                <div class="form-group">
                    <label>密码</label>
                    <input type="password" name="pwd" class="form-control" >
                </div>

                <button type="button" class="btn btn-primary">登录</button>
                <div class="alert text-danger  <c:if test="${msg==null}">d-none</c:if>">${msg}</div>
            </form>
        </div>
        <script src="/statics/js/jquery.js" type="text/javascript"></script>
        <script src="/statics/js/bootstrap.js" type="text/javascript"></script>
        <script>
            $(document).ready(function () {
                $("button").click(function () {
                    let error = "";
                    let loginForm = $("#loginForm");
                    if (loginForm.find("input[name=loginName]").val() == "") error += "请输入登录名<br>";
                    if (loginForm.find("input[name=pwd]").val() == "") error += "请输入密码<br>";
                    if (error != "") {
                        loginForm.find("div[class~=alert]").html(error);
                        loginForm.find("div[class~=alert]").removeClass("d-none");
                    } else {
                        loginForm.find("div[class~=alert]").addClass("d-none");
                        loginForm.submit();
                    }
                });
            });
        </script>
    </body>
</html>
```

【任务 6-10】完成 LogoutServlet.java

退出功能由地址/logout 对应的 LogoutServlet 完成，其 doGet()方法销毁 session 并返回 login.jsp 页面，代码如下所示。

```java
package store.controller;
//省略包导入代码
@WebServlet("/logout")
public class LogoutServlet extends HttpServlet {
    protected void doGet(HttpServletRequest request, HttpServletResponse response) throws ServletException, IOException {
        request.getSession(true).invalidate();
        request.setAttribute("msg","退出成功");
        request.getRequestDispatcher("/login.jsp").forward(request,response);
    }
}
```

6.3.6 权限功能实现

【任务 6-11】完成权限过滤器 LoginFilter.java

权限部分要求后台页面只有登录用户才能访问，通过权限过滤器 LoginFilter 对请求资源进行判断。如果请求资源路径包含 login（login.jsp 和/login 登录验证）或者是静态资源 statics 文件夹下的内容，就不需要验证登录信息。其他资源就要验证 session 中有没有登录信息，没有就返回登录页面，有登录信息就通过请求，代码如下所示。

```java
package store.filter;
//省略包导入代码
@WebFilter("/*")
public class LoginFilter implements Filter {
    public void init(FilterConfig filterConfig) throws ServletException {}
    public void doFilter(ServletRequest servletRequest, ServletResponse servletResponse, FilterChain filterChain) throws IOException, ServletException {
        HttpServletRequest httpServletRequest=(HttpServletRequest)servletRequest;
        String uri=httpServletRequest.getRequestURI();
        HttpSession session= httpServletRequest.getSession(true);
        if(uri.contains("login")||uri.startsWith("/statics")||session.getAttribute("user")!=null){
            filterChain.doFilter(servletRequest,servletResponse);
        }else{
            servletRequest.setAttribute("msg","请先登录");
            servletRequest.getRequestDispatcher("/login.jsp").forward(servletRequest,servletResponse);
        }
    }
    public void destroy() {}
}
```

代码完成后重启服务器，未登录的用户会跳转到登录页面。

本章习题

1. 根据本章内容完成数据库创建。
2. 根据本章内容完成项目创建。
3. 根据本章内容完成实体类编写。
4. 根据本章内容完成工具类编写。
5. 根据本章内容实现登录和退出功能。
6. 根据本章内容实现权限功能。

第7章
供应商管理模块实现

本章目标

- 理解供应商管理需求
- 理解供应商数据表设计
- 理解供应商管理各功能时序图
- 掌握 Java 开发技巧
- 掌握 Bootstrap 开发技巧
- 掌握 jQuery 开发技巧

通过前一章的学习，读者已经对超市管理系统的需求有了大致的理解，并且已经完成了项目基本搭建。后续内容通过任务的方式依次完成超市管理系统的各种功能，帮助读者加深对 Bootstrap、jQuery 及 Servlet 等知识点的理解，锻炼使用编程语言解决实际问题的能力，强化基于 MVC 模式的编程思维。

7.1 查询供应商

仔细阅读"查询供应商"用例描述要求，理解时序图的含义，按要求完成编码工作。

7.1.1 任务需求

查询供应商用例描述如表 7-1 所示。

表 7-1 查询供应商用例描述

用例名称	查询供应商
编号	01
参与者	管理员
简要说明	管理员输入条件查询供应商
基本事件流	1. 管理员在查询表单中输入查询条件，单击"查询"按钮 2. 系统显示符合条件的供应商
其他事件流	无
异常事件流	返回异常页面，提示错误信息
前置条件	管理员已经登录系统
后置条件	无

查询供应商功能时序图如图 7-1 所示。

时序图中各参与者说明如下。

searchSupplierForm：视图层，放置于 supplier.jsp 里面的表单，提交方式为 post，提交地址为/supplier/list，包含控件供应商名 name。

ListSupplier：控制层，位于 store.controller 包中，是一个 Servlet。其请求地址为/supplier/list，处理来自 searchSupplierForm 的 post 请求，用于获取查询条件供应商名 name 并调用 SupplierDAO 的 findSuppliers(String name, Pager pager)方法进行分页查询，操作完成后返回供应商查询视图。当单击"上一页"或"下一页"按钮进行数据查询时，超链接发送的是 get 方式的请求，由于其处理逻辑和表单 post 方式请求完全相同，因此 ListSupplier 中 doGet()方法直接调用 doPost()方法。

SupplierDAO：数据访问模型层。其 findSuppliers(String name , Pager pager)方法进行分页查询，findSuppliers(String name)方法进行不需要分页的查询。

图 7-1 查询供应商功能时序图

/supplier.jsp：供应商管理页面的地址，请求该地址最终会返回供应商管理视图层。

在条件输入框中输入查询条件后单击"查询"按钮，分页显示满足条件的查询结果，如图 7-2 所示。

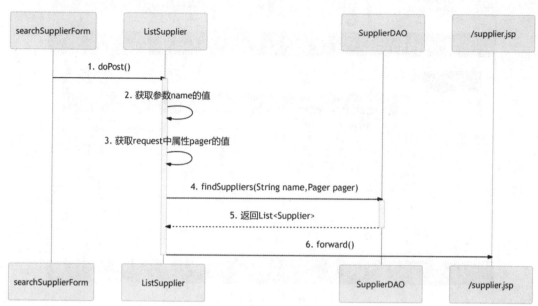

图 7-2　查询供应商页面

单击"下一页"按钮后会分页显示下一页的内容，如图 7-3 所示。

图 7-3 分页切换

7.1.2 任务实现

【任务 7-1】完成方法 SupplierDAO.findSuppliers(String name, Pager pager)

在 store.dao 包下新建 SupplierDAO 类，并在其中添加方法 findSuppliers(String name, Pager pager)。

SupplierDAO 用来封装对 t_supplier 表的操作，其 findSuppliers(String name, Pager pager) 方法进行分页查询，其中 name 为查询条件供应商名，pager 为分页帮助对象。与不需要分页的查询不同，在分页查询方法中首先通过 select count(id) as total from t_supplier where name like? 语句查询满足条件的记录总数，然后通过 select * from t_supplier where name like ? limit?,? 语句来实现分页查询，其中 limit ?,? 中的第一个占位符表示从第几条开始查询，第二个占位符表示查询最多返回几条数据。代码如下所示。

```
package store.dao;
//省略包导入代码
public class SupplierDAO {
    public List<Supplier> findSuppliers(String name, Pager pager) throws Exception {
        List<Supplier> list=new ArrayList<>();
        Connection con= null;
        PreparedStatement ps=null;
        ResultSet rs=null;
        try {
            Class.forName(DB.JDBC_DRIVER);
            con=DriverManager.getConnection(DB.JDBC_URL,DB.JDBC_USER,DB.JDBC_PASSWORD);
            ps=con.prepareStatement("select count(id) as total from t_supplier where name like ?");
```

```
                    ps.setString(1,"%"+name+"%");
                    rs=ps.executeQuery();
                    if(rs.next()){
                        pager.setTotal(rs.getInt("total"));
                    }
                    ps=con.prepareStatement("select * from t_supplier where name like? limit ?,?");
                    ps.setString(1,"%"+name+"%");
                    ps.setInt(2,(pager.getCurrentPage()-1)*pager.getPageSize());
                    ps.setInt(3,pager.getPageSize());
                    rs=ps.executeQuery();
                    while (rs.next()){
                        Supplier s=new Supplier();
                        s.setId(rs.getInt("id"));
                        s.setName(rs.getString("name"));
                        s.setTel(rs.getString("tel"));
                        s.setContacts(rs.getString("contacts"));
                        s.setInfo(rs.getString("info"));
                        list.add(s);
                    }
            } catch (Exception e) {
                e.printStackTrace();
                throw new Exception("数据库异常:"+e.getMessage());
            }finally {
                if(rs!=null) rs.close();
                if(ps!=null) ps.close();
                if(con!=null) con.close();
            }
            return list;
        }
    }
```

【任务 7-2】完成 ListSupplier.java

在 store.controller 包下新建 Servlet 类 ListSupplier,其请求地址为/supplier/list 。

ListSupplier 的 doPost() 方法用来处理供应商查询表单的请求,doGet()方法用来处理分页按钮的请求,由于两者的逻辑相同,因此在 doGet()方法中直接调用 doPost()方法即可。ListSupplier 从请求参数中得到查询条件 name,从请求对象的属性中得到分页对象 pager,将两者作为参数对 SupplierDAO 的 findSuppliers()方法进行调用,得到分页后的查询结果 supplierList。查询完毕后通过 request 对象的属性将 pager 和 supplierList 两个对象转发到 supplier.jsp 页面,代码如下所示。

```
package store.controller;
//省略包导入代码
@WebServlet("/supplier/list")
public class ListSupplier extends HttpServlet {
    protected void doPost(HttpServletRequest request, HttpServletResponse response) throws ServletException, IOException {
        String name = request.getParameter("name");
        if (name == null) name = "";
        try {
            Pager pager = (Pager) request.getAttribute("pager");
```

```
                List<Supplier> supplierList = new SupplierDAO().findSuppliers
(name,pager);
                request.setAttribute("supplierList", supplierList);
                request.setAttribute("pager",pager);
                request.getRequestDispatcher("/supplier.jsp").forward(request,
response);
        } catch (Exception e) {
            e.printStackTrace();
            throw new RuntimeException(e.getMessage());
        }
    }
    protected void doGet(HttpServletRequest request, HttpServletResponse
response) throws ServletException, IOException {
        doPost(request, response);
    }
}
```

【任务 7-3】完成视图部分

查询供应商表单的实现代码位于 supplier.jsp 中，以 post 方式将请求提交到/supplier/list，提交的参数 name 表示要查询的供应商名，代码如下所示。

```
<form class="form-inline" action="/supplier/list" method="post">
    <input type="text" name="name" class="form-control" placeholder="请输入查询条件">
    <button type="submit" class="btn btn-success ml-2">
        <i class="bi bi-search"></i>查询
    </button>
</form>
```

在显示查询数据时，如果数据数量偏多，一次全部显示会给用户带来不好的操作体验，例如很难找到需要的信息、页面加载慢，此时可以通过分页的方式对数据进行分割显示。在提高用户操作体验的同时，分页显示数据也可以减少服务器的数据传输量，提高服务器性能。

ListSupplier 会通过 request 作用域返回两个对象，分别是查询结果集 supplierList 和分页对象 pager，在 supplier.jsp 中通过循环 supplierList 生成结果表格，代码如下所示。

```
<table class="table table-hover table-bordered table-striped text-center">
    <thead class="thead-dark">
    <tr>
        <th scope="col">序号</th>
        <th scope="col">供应商名称</th>
        <th scope="col">联系人</th>
        <th scope="col">联系电话</th>
        <th scope="col">简介</th>
        <th scope="col">操作</th>
    </tr>
    </thead>
    <tbody>
    <c:forEach items="${requestScope.supplierList}" var="s" varStatus="vs">
        <tr>
```

```
            <td>${vs.count}</td>
            <td>${s.name}</td>
            <td>${s.contacts}</td>
            <td>${s.tel}</td>
            <td class="text-truncate" style="max-width: 150px;">${s.info}</td>
            <td>
                <div class="btn-group">
                    <a class="btn btn-sm btn-warning text-white" href="#"
                                                data-toggle="modal" data-target="#editSupplier" supplier-id="${s.id}">
                        <i class="bi-gear"></i>编辑</a>
                    <a class="btn btn-sm btn-danger" href="/supplier/delete?id=${s.id}">
                        <i class="bi-trash"></i>删除</a>
                </div>
            </td>
        </tr>
    </c:forEach>
    </tbody>
</table>
```

分页按钮部分代码如下所示，其中超链接请求地址/supplier/list 表示进行分页查询；传递参数 currentPage 表示希望跳转到第几页，从分页对象 pager 中获取；name 为输入的查询条件，从请求参数中获取。

```
<nav>
    <ul class="pagination justify-content-between">
        <li class="page-item">
            <a class="page-link rounded-pill" href="/supplier/list?currentPage=${requestScope.pager.previous}&name=${param.name}" tabindex="-1">上一页</a>
        </li>
        <li class="page-item"><a class="page-link rounded-pill">总共${requestScope.pager.pageCount}页,当前第${requestScope.pager.currentPage}页</a></li>
        <li class="page-item">
            <a class="page-link rounded-pill"href="/supplier/list?currentPage=${requestScope.pager.next}&name=${param.name}">下一页</a>
        </li>
    </ul>
</nav>
```

7.2 添加供应商

仔细阅读"添加供应商"用例描述要求，理解时序图的含义，按要求完成编码工作。

7.2.1 任务需求

添加供应商用例描述如表 7-2 所示。

表 7-2 添加供应商用例描述

用例名称	添加供应商
编号	02
参与者	管理员
简要说明	管理员通过供应商添加页面添加新的供应商
基本事件流	1. 管理员单击"添加供应商"按钮 2. 系统显示添加供应商表单 3. 管理员在表单中输入供应商名、联系人、联系电话、简介 4. 管理员单击"提交"按钮，新供应商被保存
其他事件流	在单击"提交"按钮时，如果有必填内容未填写，则提示管理员，数据不会被保存
异常事件流	返回异常页面，提示错误信息
前置条件	管理员已经登录系统
后置条件	供应商被保存到数据库

添加供应商功能时序图如图 7-4 所示。

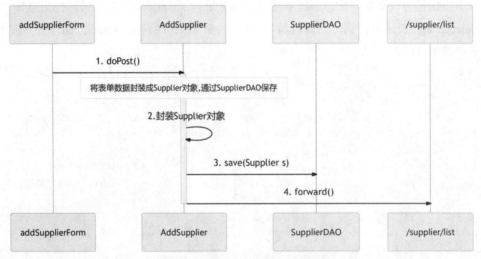

图 7-4 添加供应商功能时序图

时序图中各参与者说明如下。

addSupplierForm：视图层，放置于 supplier.jsp 中 addSupplier 模态框里面的表单，提交方式为 post，提交地址为/supplier/add，包含控件供应商名 name、联系人 contacts、联系电话 tel、简介 info。

AddSupplier：控制层，位于 store.controller 包中，是一个 Servlet。其请求地址为/supplier/add，处理来自 addSupplierForm 的 post 请求，用于封装表单数据到 Supplier 对象中并调用 SupplierDAO 的 save()方法保存数据，操作完成后返回供应商查询视图。

SupplierDAO：数据访问模型层。其 save(Supplier s)方法将一个 Supplier 对象的值通过 JDBC 添加到 t_supplier 表中。

/supplier/list：供应商查询控制层对应的地址，请求该地址最终会返回供应商查询视图层。

单击"添加供应商"按钮，弹出"添加供应商"模态框，如图 7-5 所示。

图 7-5 "添加供应商"模态框

单击"添加供应商"模态框的"提交"按钮，系统会对输入数据进行校验，如果存在必填内容没有输入，会提醒管理员输入，直到数据校验通过后表单才会被提交，校验效果如图 7-6 所示。

图 7-6 校验失败效果

7.2.2 任务实现

【任务 7-4】完成方法 SupplierDAO.save(Supplier s)

编辑 store.dao 包下的 SupplierDAO 类，添加 save(Supplier s)方法。

SupplierDAO 用来封装对 t_supplier 表的操作，其中 save(Supplier s)方法负责将传入的参数对象保存到数据库中，代码如下所示。

```
package store.dao;
//省略包导入代码
public class SupplierDAO {
    public void save(Supplier s) throws Exception {
        Connection con= null;
        PreparedStatement ps=null;
        try {
            Class.forName(DB.JDBC_DRIVER);
            con=DriverManager.getConnection(DB.JDBC_URL,DB.JDBC_USER,DB.JDBC_PASSWORD);
            ps=con.prepareStatement("insert into t_supplier value (null ,?,?,?,?)");
            ps.setString(1,s.getName());
            ps.setString(2,s.getContacts());
            ps.setString(3,s.getTel());
            ps.setString(4,s.getInfo());
            ps.executeUpdate();
        } catch (Exception e) {
            e.printStackTrace();
            throw new Exception("数据库异常:"+e.getMessage());
        }finally {
            if(ps!=null) ps.close();
            if(con!=null) con.close();
        }
    }
}
```

【任务 7-5】完成 AddSupplier.java

在 store.controller 包下新建 Servlet 类 AddSupplier，其请求地址为 /supplier/add。

AddSupplier 用来接收 addSupplierForm 表单的请求数据。它将数据封装成一个 Supplier 对象，然后调用 SupplierDAO 的 save()方法进行保存。需要注意的是，因为是添加操作，被添加的数据还没有主键，所以在封装被添加对象时不需要为 id 赋值。保存成功后转发到资源地址/supplier/list，该资源会进行一次供应商查询，代码如下所示。

```
package store.controller;
//省略包导入代码
@WebServlet("/supplier/add")
public class AddSupplier extends HttpServlet {
    protected void doPost(HttpServletRequest request, HttpServletResponse response) throws ServletException, IOException {
        Supplier s=new Supplier();
        s.setName(request.getParameter("name"));
        s.setContacts(request.getParameter("contacts"));
        s.setTel(request.getParameter("tel"));
        s.setInfo(request.getParameter("info"));
        try {
            new SupplierDAO().save(s);
            request.getRequestDispatcher("/supplier/list").forward(request,
```

```
response);
            } catch (Exception e) {
                e.printStackTrace();
                throw new RuntimeException(e.getMessage());
            }
        }
    }
```

【任务 7-6】完成视图部分

添加供应商信息输入界面通过 supplier.jsp 页面中的"添加供应商"按钮触发模态框生成，代码如下所示。

```
<button type="button" class="btn btn-primary" data-toggle="modal" data-target="#addSupplier"><i class="bi-plus"></i>添加供应商</button>
```

代码中 data-toggle="modal"用来指定 Bootstrap 的触发行为是模态框，data-target="#addSupplier"用来指定要显示的模态框组件 id，对应下面代码中的 id="addSupplier"。

```html
<!-- 添加供应商 Modal -->
<div class="modal fade" id="addSupplier" tabindex="-1">
    <div class="modal-dialog">
        <div class="modal-content">
            <div class="modal-header">
                <h5 class="modal-title">添加供应商</h5>
                <button type="button" class="close" data-dismiss="modal">
                    <span>&times;</span>
                </button>
            </div>
            <div class="modal-body">
                <form id="addSupplierForm" action="/supplier/add" method="post">
                    <div class="form-group row">
                        <label class="col-sm-2 col-form-label">供应商</label>
                        <div class="col-sm-10">
                            <input type="text" name="name" class="form-control"placeholder="供应商名">
                        </div>
                    </div>
                    <div class="form-group row">
                        <label class="col-sm-2 col-form-label">联系人</label>
                        <div class="col-sm-10">
                            <input type="text" name="contacts" class="form-control" placeholder="联系人">
                        </div>
                    </div>

                    <div class="form-group row">
                        <label class="col-sm-2 col-form-label">电话</label>
                        <div class="col-sm-10">
                            <input type="text" name="tel" class="form-control" placeholder="电话">
                        </div>
                    </div>
                    <div class="form-group row">
```

```
                                <label class="col-sm-2 col-form-label">简介</label>
                                <div class="col-sm-10">
                                    <textarea name="info" class="form-control" rows="4"></textarea>
                                </div>
                            </div>
                            <div class="text-danger d-none alert">
                            </div>
                        </form>
                    </div>
                    <div class="modal-footer">
                        <button type="button" class="btn btn-secondary" data-dismiss="modal">关闭</button>
                        <button id="addButton" type="button" class="btn btn-primary">提交</button>
                    </div>
                </div>
            </div>
        </div>
```

代码 `<button id="addButton" type="button" class="btn btn-primary">` 中为"提交"按钮设置了 id 值, 通过 jQuery 代码为该按钮绑定了 click 事件处理代码, 当"提交"按钮被单击时通过 CSS 选择器找到 id="addSupplierForm"表单里面需要进行非空验证的控件进行数据验证。

`$("#addButton").click(function () { ... })`用来为前文中 id="addButton"的按钮绑定 click 事件处理代码。事件处理代码中 `let addForm = $("#addSupplierForm");` 通过 jQuery 选择器返回 id="addSupplierForm" 的表单元素, 通过该元素对象调用 find() 方法可以查找元素内部的指定子元素对象。例如, 代码 `addForm.find("input[name=name]")` 中该元素对象通过属性选择器找到子元素中属性 name="name" 的 input 元素, 通过 input 元素的 val()方法可以取到输入框中的值, 如果没有输入, 就在错误提示中进行提示。代码如下所示。

```
$(document).ready(function () {
    $("#addButton").click(function () {
        let error = "";
        let addForm = $("#addSupplierForm");
        if (addForm.find("input[name=name]").val() == "") {
            error += "请输入供应商名<br>";
        }
        if (addForm.find("input[name=contacts]").val() == "") {
            error += "请输入联系人<br>";
        }
        if (addForm.find("input[name=tel]").val() == "") {
            error += "请输入联系电话<br>";
        }
        if (error != "") {
            addForm.find("div[class~=alert]").html(error);
            addForm.find("div[class~=alert]").removeClass("d-none");
        } else {
            addForm.find("div[class~=alert]").addClass("d-none");
            addForm.submit();
```

            }
        });
}

7.3 删除供应商

仔细阅读"删除供应商"用例描述要求，理解时序图的含义，按要求完成编码工作。

7.3.1 任务需求

删除供应商用例描述如表 7-3 所示。

表 7-3 删除供应商用例描述

用例名称	删除供应商
编号	03
参与者	管理员
简要说明	管理员删除指定供应商
基本事件流	1. 管理员在查询表单中输入查询条件，单击"查询"按钮 2. 系统显示符合条件的供应商 3. 管理员单击指定供应商的"删除"超链接 4. 系统删除相应的供应商记录 5. 系统返回供应商查询页面
其他事件流	无
异常事件流	返回异常页面，提示错误信息
前置条件	管理员已经登录系统
后置条件	相应的供应商记录从数据库中删除

删除供应商功能时序图如图 7-7 所示。

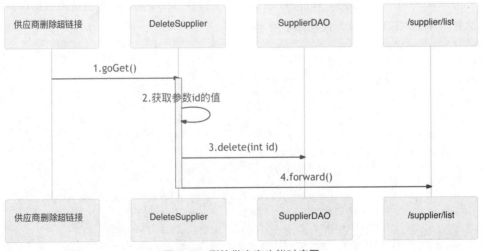

图 7-7 删除供应商功能时序图

时序图中各参与者说明如下。

供应商删除超链接：视图层，供应商查询结果页面列表中的每一条供应商记录都有一个对应的"删除"超链接，供应商删除超链接会向后台传递要删除的供应商主键。

DeleteSupplier：控制层，位于 store.controller 包中，是一个 Servlet。其请求地址为/supplier/delete，处理来自供应商删除超链接的 get 请求，从请求参数中获取需要删除的供应商主键并调用 SupplierDAO 的 delete()方法删除数据，操作完成后返回供应商查询视图。

SupplierDAO：数据访问模型层。其 delete()方法通过 JDBC 将 t_supplier 表中对应主键的记录删除。

/supplier/list：供应商查询控制层对应的地址，请求该地址最终会返回供应商查询视图层。

在供应商查询结果列表中的每一条供应商记录都有对应的"删除"超链接，单击"删除"超链接可以从数据库中删除其对应的供应商记录，如图 7-8 所示。

序号	供应商名称	联系人	联系电话	简介	操作
1	测试供应商2	联系人2	13980000002	介绍2	编辑 删除
2	测试供应商3	联系人3	13980000003	介绍3	编辑 删除
3	测试供应商4	联系人4	13980000004	介绍4	编辑 删除
4	测试供应商9	联系人9	13980000009	介绍9	编辑 删除
5	测试供应商11	联系人11	13980000011	介绍11	编辑 删除

图 7-8 删除供应商界面

7.3.2 任务实现

【任务 7-7】完成方法 SupplierDAO.delete(int id)

编辑 store.dao 下的 SupplierDAO 类，添加 delete(int id)方法。

SupplierDAO 用来封装对 t_supplier 表的操作，其中 delete(int id)方法负责根据传入的主键删除数据库中对应的记录，代码如下所示。

```
package store.dao;
//省略包导入代码
public class SupplierDAO {
    public void delete(int id) throws Exception {
        Connection con= null;
        PreparedStatement ps=null;
        try {
            Class.forName(DB.JDBC_DRIVER);
            con=DriverManager.getConnection(DB.JDBC_URL,DB.JDBC_USER, DB.JDBC_PASSWORD);
            ps=con.prepareStatement("delete from t_supplier where id=?");
            ps.setInt(1,id);
            ps.executeUpdate();
        } catch (Exception e) {
            e.printStackTrace();
```

```
            throw new Exception("数据库异常:"+e.getMessage());
        }finally {
            if(ps!=null) ps.close();
            if(con!=null) con.close();
        }
    }
}
```

【任务 7-8】完成 DeleteSupplier.java

在 store.controller 包下新建 Servlet 类 DeleteSupplier,其请求地址为/supplier/delete。

DeleteSupplier 用来接收供应商删除超链接的请求,从请求参数中获取需要删除的供应商主键,然后将主键作为参数调用 SupplierDAO 的 delete(int id)方法。记录删除成功后转发到资源地址/supplier/list,该资源会进行一次无参数的供应商查询。代码如下所示。

```
package store.controller;
//省略包导入代码
@WebServlet("/supplier/delete")
public class DeleteSupplier extends HttpServlet {
    protected void doGet(HttpServletRequest request, HttpServletResponse response) throws ServletException, IOException {
        int id=Integer.parseInt(request.getParameter("id"));
        try {
            new SupplierDAO().delete(id);
            request.getRequestDispatcher("/supplier/list").forward(request, response);
        } catch (Exception e) {
            e.printStackTrace();
            throw new RuntimeException(e.getMessage());
        }
    }
}
```

【任务 7-9】完成视图部分

supplier.jsp 中包含供应商查询结果显示代码,通过 JSTL 的 forEach 标签循环查询结果,每次循环生成一行结果,代码如下所示。

```
<c:forEach items="${requestScope.supplierList}" var="s" varStatus="vs">
    <tr>
        <td>${vs.count}</td>
        <td>${s.name}</td>
        <td>${s.contacts}</td>
        <td>${s.tel}</td>
        <td class="text-truncate" style="max-width: 150px;">${s.info}</td>
        <td>
            <div class="btn-group">
                <a class="btn btn-sm btn-warning text-white" href="#"
                                    data-toggle="modal" data-target="#editSupplier" supplier-id="${s.id}">
                    <i class="bi-gear"></i>编辑</a>
                <a class="btn btn-sm btn-danger" href="/supplier/delete?id=${s.id}">
                    <i class="bi-trash"></i>删除</a>
            </div>
```

```
                </td>
            </tr>
</c:forEach>
```

其中生成删除超链接的代码 href="/supplier/delete?id=${s.id}" 通过 EL 表达式从当前循环的供应商对象中取出主键并赋给超链接的参数 id，当该超链接被单击时，其对应的主键就会被传递给后台。

7.4 修改供应商

仔细阅读"修改供应商"用例描述要求，理解时序图的含义，按要求完成编码工作。

7.4.1 任务需求

修改供应商用例描述如表 7-4 所示。

表 7-4 修改供应商用例描述

用例名称	修改供应商
编号	04
参与者	管理员
简要说明	管理员对供应商信息进行编辑后将数据更新到数据库
基本事件流	1. 管理员在供应商查询结果页面单击需要更新的供应商对应的"编辑"超链接 2. 系统显示供应商编辑界面 3. 管理员在编辑界面编辑数据后单击"提交"按钮 4. 系统保存修改后的数据 5. 系统返回供应商查询页面
其他事件流	无
异常事件流	返回异常页面，提示错误信息
前置条件	管理员已经登录系统
后置条件	供应商数据被更新到数据库

修改供应商功能时序图如图 7-9 所示。

时序图中各参与者说明如下。

supplier.jsp：视图层。供应商查询结果页面列表中的每一条供应商记录都有一个对应的"编辑"超链接，单击"编辑"超链接会发送一个 AJAX 请求，将需要更新的供应商主键传给后台。

GetSupplier：控制层，位于 store.controller 包中，是一个 Servlet。其请求地址为/supplier，处理来自 supplier.jsp 的异步 get 请求，用于获取请求参数中的供应商主键并调用 SupplierDAO 的 findById(int id)方法查询供应商信息，操作完成后以 JSON 格式返回供应商信息。

EditSupplier：控制层，位于 store.controller 包中，是一个 Servlet。其请求地址为/supplier/edit，处理来自 supplier.jsp 的 post 请求，用于获取请求参数中的供应商修改信息并将其封装成 Supplier 对象，调用 SupplierDAO 的 update(Supplier s) 方法更新供应商信息，操作完成后返回供应商信息查询页面。

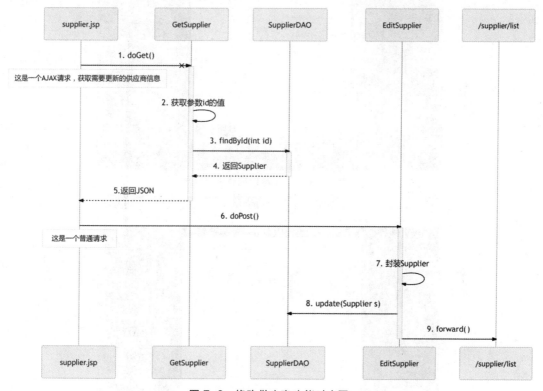

图 7-9　修改供应商功能时序图

/supplier/list：供应商查询控制层对应的地址，请求该地址最终会返回供应商查询视图层。

SupplierDAO：数据访问模型层。其 findById(int id)方法根据 id 查找对应的供应商信息并返回。

【注意】

供应商信息修改需要两个步骤。第一步：通过 AJAX 请求查询需要修改的供应商信息，服务器以 JSON 格式返回数据，前端 jQuery 代码接收到数据后填充供应商信息编辑界面。第二步：管理员修改完成后单击"提交"按钮，通过表单以同步消息发送修改后的数据给服务器保存。

在供应商查询结果页面单击供应商对应的"编辑"超链接，弹出供应商信息编辑界面，如图 7-10 所示。

编辑供应商信息后单击"提交"按钮，如果信息输入不完整，会给出相应提示，如图 7-11 所示。

信息输入无误并提交后，数据被保存到数据库中，系统返回供应商查询结果页面，显示修改后的数据，如图 7-12 所示。

图 7-10 供应商信息编辑界面

图 7-11 修改供应商失败

图 7-12 修改供应商成功

7.4.2 任务实现

【任务 7-10】完成方法 SupplierDAO.findById(int id)

编辑 store.dao 包下的 SupplierDAO，并添加 findById(int id)方法。

SupplierDAO 用来封装对 t_supplier 表的操作。findById(int id)方法根据参数 id 查询供应商表 t_supplier，将结果封装成 Supplier 返回，代码如下所示。

```
package store.dao;
//省略包导入代码
public class SupplierDAO {
```

```
        public Supplier findById(int id) throws Exception {
            Supplier s=null;
            Connection con= null;
            PreparedStatement ps=null;
            ResultSet rs=null;
            try {
                Class.forName(DB.JDBC_DRIVER);
                con=DriverManager.getConnection(DB.JDBC_URL,DB.JDBC_USER,
DB.JDBC_PASSWORD);
                ps=con.prepareStatement("select * from t_supplier where id=? ");
                ps.setInt(1,id);
                rs=ps.executeQuery();
                if (rs.next()){
                    s=new Supplier();
                    s.setId(rs.getInt("id"));
                    s.setName(rs.getString("name"));
                    s.setTel(rs.getString("tel"));
                    s.setContacts(rs.getString("contacts"));
                    s.setInfo(rs.getString("info"));
                }
            } catch (Exception e) {
                e.printStackTrace();
                throw new Exception("数据库异常:"+e.getMessage());
            }finally {
                if(rs!=null) rs.close();
                if(ps!=null) ps.close();
                if(con!=null) con.close();
            }
            return s;
        }
    }
```

【任务 7-11】完成方法 SupplierDAO.update(Supplier s)

编辑 store.dao 包下的 SupplierDAO 类，添加 update(Supplier s)方法。SupplierDAO 用来封装对 t_supplier 表的操作。update(Supplier s)方法根据参数 s 的 id 将其属性值更新到数据库对应的记录中，代码如下所示。

```
package store.dao;
//省略包导入代码
public class SupplierDAO {
    public void update(Supplier s) throws Exception {
        Connection con= null;
        PreparedStatement ps=null;
        try {
            Class.forName(DB.JDBC_DRIVER);
            con=DriverManager.getConnection(DB.JDBC_URL,DB.JDBC_USER,
DB.JDBC_PASSWORD);
            ps=con.prepareStatement("update t_supplier set name=?,contacts=?,
tel=?,info=? where id=?");
            ps.setString(1,s.getName());
            ps.setString(2,s.getContacts());
            ps.setString(3,s.getTel());
            ps.setString(4,s.getInfo());
            ps.setInt(5,s.getId());
            ps.executeUpdate();
```

```
            } catch (Exception e) {
                e.printStackTrace();
                throw new Exception("数据库异常:"+e.getMessage());
            }finally {
                if(ps!=null) ps.close();
                if(con!=null) con.close();
            }
        }
    }
```

【任务 7-12】 完成 GetSupplier.java

在 store.controller 包下新建 Servlet 类 GetSupplier,其请求地址为/supplier。

GetSupplier 的 doGet()方法接收前端的 AJAX 请求,通过请求获取参数供应商主键,然后通过 SupplierDAO 查询对应的供应商信息,最后以 JSON 格式返回数据。GetSupplier 数据处理过程跟普通请求一样,重点在于返回数据的方式变化。传统同步请求 Servlet 通过重定向或者转发的方式返回一个新的视图,但是在异步请求的情况下,页面中的视图只是局部更新,所以 Servlet 不用返回新的视图,只需要返回页面需要的数据即可。因此 Servlet 将数据转换成 JSON 对象,然后通过输出流将对象以 JSON 数据返回。代码如下所示。

```
package store.controller;
//省略包导入代码
@WebServlet("/supplier")
public class GetSupplier extends HttpServlet {
    protected void doGet(HttpServletRequest request, HttpServletResponse response) throws ServletException, IOException {
        int id=Integer.parseInt(request.getParameter("id"));
        JSONObject jsonObject=new JSONObject();
        try {
            Supplier s=new SupplierDAO().findById(id);
            jsonObject.put("success",true);
            jsonObject.put("data",s);
        } catch (Exception e) {
            e.printStackTrace();
            jsonObject.put("success",false);
            jsonObject.put("msg",e.getMessage());
        }finally{
            response.setContentType("application/x-json");
            response.getWriter().write(jsonObject.toJSONString());
            response.getWriter().flush();
            response.getWriter().close();
        }
    }
}
```

response.setContentType("application/x-json")将响应数据类型设置为 JSON 格式,页面中 jQuery 的 getJSON()方法通过返回值中的 data 属性就可以直接访问供应商的属性值。

【任务 7-13】 完成 EditSupplier.java

在 store.controller 包下新建 Servlet 类 EditSupplier,其请求地址为/supplier/edit。

EditSupplier 接收供应商更新表单的 post 请求,从请求参数中获取更新数据并封装成 Supplier 对象,然后调用 SupplierDAO 的 update(Supplier s)方法更新数据,完成后转发到资源路径

/supplier/list，该资源路径会返回供应商查询结果页面。代码如下所示。

```java
package store.controller;
//省略包导入代码
@WebServlet("/supplier/edit")
public class EditSupplier extends HttpServlet {
    protected void doPost(HttpServletRequest request, HttpServletResponse response) throws ServletException, IOException {
        Supplier s=new Supplier();
        s.setId(Integer.parseInt(request.getParameter("id")));
        s.setName(request.getParameter("name"));
        s.setContacts(request.getParameter("contacts"));
        s.setTel(request.getParameter("tel"));
        s.setInfo(request.getParameter("info"));
        try {
            new SupplierDAO().update(s);
            request.getRequestDispatcher("/supplier/list").forward(request, response);
        } catch (Exception e) {
            e.printStackTrace();
            throw new RuntimeException(e.getMessage());
        }
    }
}
```

【任务 7-14】完成视图部分

供应商信息编辑界面通过"编辑"超链接触发编辑模态框。表单提交前需要进行数据验证，验证通过后通过 jQuery 代码提交，因此"提交"按钮的类型是 button，而不是 submit，代码如下所示。

```html
<div class="modal fade" id="editSupplier" tabindex="-1">
    <div class="modal-dialog">
        <div class="modal-content">
            <div class="modal-header">
                <h5 class="modal-title">编辑供应商</h5>
                <button type="button" class="close" data-dismiss="modal">
                    <span>&times;</span>
                </button>
            </div>
            <div class="modal-body">
                <form id="editSupplierForm" action="/supplier/edit" method="post">
                    <div class="form-group row">
                        <label class="col-sm-2 col-form-label">供应商</label>
                        <div class="col-sm-10">
                            <input type="hidden" name="id">

                            <input type="text" name="name" class="form-control" placeholder="供应商名">
                        </div>
                    </div>
                    <div class="form-group row">
                        <label class="col-sm-2 col-form-label">联系人</label>
```

```html
                                <div class="col-sm-10">
                                    <input type="text" name="contacts" class="form-control" placeholder="联系人">
                                </div>
                            </div>

                            <div class="form-group row">
                                <label class="col-sm-2 col-form-label">联系电话</label>
                                <div class="col-sm-10">
                                    <input type="text" name="tel" class="form-control" placeholder="联系电话">
                                </div>
                            </div>
                            <div class="form-group row">
                                <label class="col-sm-2 col-form-label">简介</label>
                                <div class="col-sm-10">
                                    <textarea name="info" class="form-control" rows="4"></textarea>
                                </div>
                            </div>
                            <div class="text-danger d-none alert">
                            </div>
                        </form>
                    </div>
                    <div class="modal-footer">
                        <button type="button" class="btn btn-secondary" data-dismiss="modal">关闭</button>
                        <button id="editButton" type="button" class="btn btn-primary">提交</button>
                    </div>
                </div>
            </div>
        </div>
```

在实现供应商查询结果页面时循环输出表格的行数据，每一行都有对应的"编辑"超链接，每个超链接都含有一个自定义属性 supplier-id，该属性的值为对应的供应商主键，"编辑"超链接代码如下。

```html
<a class="btn btn-sm btn-warning text-white" href="#" supplier-id="${s.id}"
                    data-toggle="modal" data-target="#editSupplier" >
    <i class="bi-gear"></i>编辑</a>
```

通过 jQuery 代码 $("a[supplier-id]").click() 可以选中页面中所有的"编辑"超链接，并且为它们绑定单击事件。在单击事件处理代码中通过$.getJSON()向资源/supplier 发送 AJAX 请求，在接收到返回值后使用返回的 JSON 数据为模态框中的控件赋值。代码如下所示。

```html
<script>
    $(document).ready(function () {
        $("a[supplier-id]").click(function () {
            $.getJSON("/supplier", {id: $(this).attr("supplier-id")}, function (result) {
                $("#editSupplierForm").find("input[name=name]").val
```

```
(result.data.name);
                    $("#editSupplierForm").find("input[name=contacts]").val
(result.data.contacts);
                    $("#editSupplierForm").find("input[name=tel]").val(result.
data.tel);
                    $("#editSupplierForm").find("input[name=id]").val(result.
data.id);
                    $("#editSupplierForm").find("textarea[name=info]").val
(result.data.info);
                });
            });
        });
    </script>
```

修改完成后单击"提交"按钮,需要对输入的数据进行数据校验。通过 jQuery 代码 $("#editButton").click()为"提交"按钮绑定单击事件,在事件处理代码中依次对必填数据进行校验,如果验证失败,则通过 jQuery 找到表单中的警告信息显示层,并在其中显示信息。验证通过则提交表单。代码如下所示。

```
<script>
$(document).ready(function () {
    $("#editButton").click(function () {
        let error = "";
        let editForm = $("#editSupplierForm");
        if (editForm.find("input[name=name]").val() == "") {
            error += "请输入供应商名<br>";
        }
        if (editForm.find("input[name=contacts]").val() == "") {
            error += "请输入联系人<br>";
        }
        if (editForm.find("input[name=tel]").val() == "") {
            error += "请输入联系电话<br>";
        }
        if (error != "") {
            editForm.find("div[class~=alert]").html(error);
            editForm.find("div[class~=alert]").removeClass("d-none");
        } else {
            editForm.find("div[class~=alert]").addClass("d-none");
            editForm.submit();
        }
    });
});
</script>
```

本章习题

1. 结合项目代码,实现查询供应商功能。
2. 结合项目代码,实现添加供应商功能。
3. 结合项目代码,实现删除供应商功能。
4. 结合项目代码,实现修改供应商功能。

第8章
分类管理模块实现

本章目标

- 理解分类管理需求
- 理解分类数据表设计
- 理解分类管理各功能时序图
- 掌握 Java 开发技巧
- 掌握 Bootstrap 开发技巧
- 掌握 jQuery 开发技巧

通过前一章的学习，读者已经实现了供应商管理模块的功能。本章通过实现超市管理系统的分类管理模块功能，帮助读者加深对 Bootstrap、jQuery 及 Servlet 等知识点的理解，锻炼使用编程语言解决实际问题的能力，强化基于 MVC 模式的编程思维。

8.1 查询分类

仔细阅读"查询分类"用例描述要求，理解时序图的含义，按要求完成编码工作。

8.1.1 任务需求

查询分类用例描述如表 8-1 所示。

表 8-1 查询分类用例描述

用例名称	查询分类
编号	05
参与者	管理员
简要说明	管理员输入条件查询商品分类
基本事件流	1. 管理员在查询表单中输入查询条件，单击"查询"按钮 2. 系统显示符合条件的分类
其他事件流	无
异常事件流	返回异常页面，提示错误信息
前置条件	管理员已经登录系统
后置条件	无

查询分类功能时序图如图 8-1 所示。

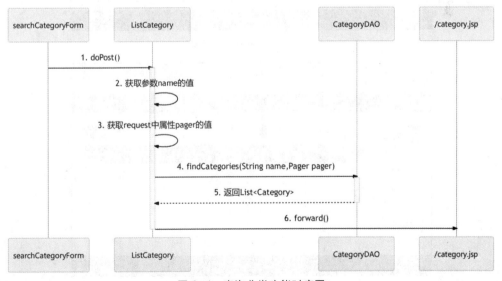

图 8-1 查询分类功能时序图

时序图中各参与者说明如下。

searchCategoryForm：视图层，放置于 category.jsp 里面的表单，提交方式为 post，提交地址为/category/list，包含控件分类名 name。

ListCategory：控制层，位于 store.controller 包中，是一个 Servlet。其请求地址为/category/list，处理来自 searchCategoryForm 的 post 请求，用于获取查询条件分类名 name 并调用 CategoryDAO 的 findCategories(String name, Pager pager)方法进行分页查询，操作完成后返回分类查询视图。当单击"上一页"或"下一页"按钮进行数据查询时，超链接发送的是 get 方式的请求，由于其处理逻辑和表单 post 方式请求完全相同，因此 ListCategory 中 doGet()方法直接调用 doPost()方法。

CategoryDAO：数据访问模型层。其 findCategories(String name, Pager pager)方法进行分页查询，findCategories(String name)方法进行不需要分页的查询。

/category.jsp：分类管理页面的地址，请求该地址最终会返回分类管理视图层。

在分类管理主页面的条件输入框中输入查询条件并单击"查询"按钮，分页显示满足条件的查询结果，如图 8-2 和图 8-3 所示。

图 8-2　分类管理主页面

图 8-3　分类查询结果

8.1.2 任务实现

【任务 8-1】完成方法 CategoryDAO.findCategories(String name, Pager pager)

在 store.dao 包下新建 CategoryDAO 类,并在其中添加方法 public List findCategories(String name, Pager pager)。

CategoryDAO 用来封装对 t_category 表的操作,其 findCategories(String name, Pager pager) 方法用于进行分页查询,其中 name 为查询条件分类名,pager 为分页帮助对象。与不需要分页的查询不同,在分页查询方法中首先通过 select count(id) as total from t_category where name like? 语句查询满足条件的记录总数,然后通过 select * from t_category where name like? limit?,? 语句来实现分页查询,其中 limit?,? 中的第一个占位符表示从第几条开始查,第二个占位符表示查询最多返回几条数据,代码如下所示。

```java
package store.dao;
//省略包导入代码
public class CategoryDAO {
    public List<Category> findCategories(String name, Pager pager) throws Exception {
        List<Category> list=new ArrayList<>();
        Connection con= null;
        PreparedStatement ps=null;
        ResultSet rs=null;
        try {
            Class.forName(DB.JDBC_DRIVER);
            con=DriverManager.getConnection(DB.JDBC_URL,DB.JDBC_USER,DB.JDBC_PASSWORD);
            ps=con.prepareStatement("select count(id) as total from t_category where name like ?");
            ps.setString(1,"%"+name+"%");
            rs=ps.executeQuery();
            if(rs.next()){
                pager.setTotal(rs.getInt("total"));
            }
            ps=con.prepareStatement("select * from t_category where name like ? limit ?,?");
            ps.setString(1,"%"+name+"%");
            ps.setInt(2,(pager.getCurrentPage()-1)*pager.getPageSize());
            ps.setInt(3,pager.getPageSize());
            rs=ps.executeQuery();
            while (rs.next()){
                Category c=new Category();
                c.setId(rs.getInt("id"));
                c.setName(rs.getString("name"));
                c.setDescription(rs.getString("description"));
                list.add(c);
            }
        } catch (Exception e) {
            e.printStackTrace();
            throw new Exception("数据库异常:"+e.getMessage());
```

```
        }finally {
            if(rs!=null) rs.close();
            if(ps!=null) ps.close();
            if(con!=null) con.close();
        }
        return list;
    }
}
```

【任务 8-2】完成 ListCategory.java

在 store.controller 包下新建 Servlet 类 ListCategory,其请求地址为/category/list。

ListCategory 的 doPost()方法用来处理分类查询表单的请求,doGet()方法用来处理分页按钮的请求,由于两者的逻辑相同,因此在 doGet()方法中直接调用 doPost()方法即可。ListCategory 从请求参数中得到查询条件 name,从请求对象的属性中得到分页对象 pager,将两者作为参数对 CategoryDAO 的 findCategories()方法进行调用,得到分页后的查询结果 categoryList。查询完毕后通过 request 对象的属性将 pager 和 categoryList 两个对象转发到 category.jsp 页面,代码如下所示。

```
package store.controller;
//省略包导入代码
@WebServlet("/category/list")
public class ListCategory extends HttpServlet {
    protected void doPost(HttpServletRequest request, HttpServletResponse response) throws ServletException, IOException {
        String name=request.getParameter("name");
        if(name==null) name="";
        try {
            Pager pager = (Pager) request.getAttribute("pager");
            List<Category> categoryList=new CategoryDAO().findCategories(name,pager);
            request.setAttribute("categoryList",categoryList);
            request.getRequestDispatcher("/category.jsp").forward(request,response);
        } catch (Exception e) {
            e.printStackTrace();
            throw new RuntimeException(e.getMessage());
        }
    }
    protected void doGet(HttpServletRequest request, HttpServletResponse response) throws ServletException, IOException {
        doPost(request,response);
    }
}
```

【任务 8-3】完成视图部分

查询分类表单的实现代码 searchCategoryForm 位于 category.jsp 中,以 post 方式提交到/category/list,提交参数 name 表示要查询的分类名,代码如下所示。

```
<form class="form-inline" action="/category/list" method="post">
    <input type="text" name="name" class="form-control" placeholder="请输入查询条件">
    <button type="submit" class="btn btn-success ml-2"><i class="bi-search">
```

```
</i>查询
    </button>
</form>
```

ListCategory 会通过 request 作用域返回两个对象，分别是分页查询结果集 categoryList 和分页对象 pager，在 category.jsp 中通过循环 categoryList 生成结果表格，代码如下所示。

```
<table class="table table-hover table-bordered table-striped text-center">
    <thead class="thead-dark">
    <tr>
        <th scope="col">序号</th>
        <th scope="col">分类名</th>
        <th scope="col">简介</th>
        <th scope="col">操作</th>
    </tr>
    </thead>
    <tbody>
    <c:forEach items="${requestScope.categoryList}" var="c" varStatus="vs">
        <tr>
            <td>${vs.count}</td>
            <td>${c.name}</td>
            <td class="text-truncate" style="max-width: 150px;">${c.description}</td>
            <td>
                <div class="btn-group">
                    <a class="btn btn-sm btn-warning text-white" href="#"
                       data-toggle="modal" data-target="#editCategory" category-id="${c.id}">
                        <i class="bi-gear"></i>编辑</a>
                    <a class="btn btn-sm btn-danger" href="/category/delete?id=${c.id}">
                        <i class="bi-trash"></i>删除</a>
                </div>
            </td>
        </tr>
    </c:forEach>
    </tbody>
</table>
```

分页按钮部分代码如下所示，其中超链接请求/category/list 进行分页查询；传递参数 currentPage 表示希望跳转到第几页，从分页对象 pager 中获取；name 为输入的查询条件，从请求参数中获取。

```
<nav>
    <ul class="pagination justify-content-between">
        <li class="page-item">
            <a class="page-link rounded-pill" href="/category/list?currentPage=${requestScope.pager.previous}&name=${param.name}" tabindex="-1">上一页</a>
        </li>
        <li class="page-item"><a class="page-link rounded-pill">总共${requestScope.pager.pageCount}页，当前第${requestScope.pager.currentPage}页</a></li>
        <li class="page-item">
```

```
                <a class="page-link rounded-pill"href="/category/list?currentPage=
${requestScope.pager.next}&name=${param.name}">下一页</a>
            </li>
        </ul>
    </nav>
```

8.2 添加分类

仔细阅读"添加分类"用例描述要求,理解时序图的含义,按要求完成编码工作。

8.2.1 任务需求

添加分类用例描述如表 8-2 所示。

表 8-2 添加分类用例描述

用例名称	添加分类
编号	06
参与者	管理员
简要说明	管理员通过分类添加页面添加新的商品分类
基本事件流	1. 管理员单击"添加分类"按钮 2. 系统显示添加分类表单 3. 管理员在表单中输入分类名、分类介绍 4. 管理员单击"提交"按钮,新分类被保存
其他事件流	在单击"提交"按钮时,如果有必填内容未填写,则提示管理员,数据不会被保存
异常事件流	返回异常页面,提示错误信息
前置条件	管理员已经登录系统
后置条件	分类被保存到数据库

添加分类功能时序图如图 8-4 所示。

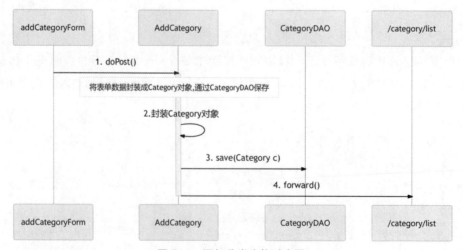

图 8-4 添加分类功能时序图

时序图中各参与者说明如下。

addCategoryForm：视图层，放置于 category.jsp 中 addCategory 模态框里面的表单，提交方式为 post，提交地址为/category/add，包含控件分类名 name、分类描述 description。

AddCategory：控制层，位于 store.controller 包中，是一个 Servlet。其请求地址为/category/add，处理来自 addCategoryForm 的 post 请求，用于封装表单数据到 Category 对象中并调用 CategoryDAO 的 save()方法保存数据，操作完成后返回分类查询视图。

CategoryDAO：数据访问模型层。其 save(Category c)方法将一个 Category 对象的值通过 JDBC 添加到 t_category 表中。

/category/list：分类查询控制层对应的地址。请求该地址最终会返回分类查询视图层。

单击"添加分类"按钮，弹出"添加分类"模态框，如图 8-5 所示。

图 8-5 "添加分类"模态框

单击"添加分类"模态框中的"提交"按钮，系统会对输入数据进行校验，如果存在必填内容没有输入，会提醒管理员输入，直到数据校验通过后表单才会提交，校验效果如图 8-6 所示。

图 8-6 添加分类校验失败

分类信息输入无误后,单击"提交"按钮,分类信息被保存到数据库,系统返回分类查询页面,显示刚刚添加的分类信息,如图 8-7 所示。

图 8-7 分类添加成功

8.2.2 任务实现

【任务 8-4】完成方法 CategoryDAO.save(Category c)

编辑 store.controller 包下的 CategoryDAO 类,添加 save(Category c)方法。

CategoryDAO 用来封装对 t_category 表的操作,其中 save(Category c)方法负责将传入的参数对象保存到数据库中,代码如下所示。

```java
package store.dao;
//省略包导入代码
public class CategoryDAO {
    public void save(Category c) throws Exception {
        Connection con= null;
        PreparedStatement ps=null;
        try {
            Class.forName(DB.JDBC_DRIVER);
            con=DriverManager.getConnection(DB.JDBC_URL,DB.JDBC_USER,DB.JDBC_PASSWORD);
            ps=con.prepareStatement("insert into t_category value (null ,?,?)");
            ps.setString(1,c.getName());
            ps.setString(2,c.getDescription());
            ps.executeUpdate();
        } catch (Exception e) {
            e.printStackTrace();
            throw new Exception("数据库异常:"+e.getMessage());
        }finally {
            if(ps!=null) ps.close();
            if(con!=null) con.close();
        }
    }
}
```

【任务 8-5】完成 AddCategory.java

在 store.controller 包下新建 Servlet 类 AddCategory，其请求地址为/category/add。

AddCategory 用来接收 addCategoryForm 表单的请求数据。它将数据封装成一个 Category 对象，然后调用 CategoryDAO 的 save()方法进行保存。需要注意的是，因为是添加操作，被添加的数据还没有主键，所以在封装被添加对象时不需要为 id 赋值。保存成功后转发到资源地址/category/list，该资源会进行一次分类查询，代码如下所示。

```java
package store.controller;
//省略包导入代码
@WebServlet("/category/add")
public class AddCategory extends HttpServlet {
    protected void doPost(HttpServletRequest request, HttpServletResponse response) throws ServletException, IOException {
        Category c=new Category();
        c.setName(request.getParameter("name"));
        c.setDescription(request.getParameter("description"));
        try {
            new CategoryDAO().save(c);
            request.getRequestDispatcher("/category/list").forward(request, response);
        } catch (Exception e) {
            e.printStackTrace();
            throw new RuntimeException(e.getMessage());
        }
    }
}
```

【任务 8-6】完成视图部分

添加分类信息输入界面通过 category.jsp 页面中的"添加分类"按钮触发模态框生成，代码如下所示。

```html
<button type="button" class="btn btn-primary" data-toggle="modal" data-target="#addCategory"><i class="bi-plus"></i>添加分类</button>
```

代码中 data-toggle="modal"用来指定 Bootstrap 的触发行为是模态框，data-target="#addCategory" 用来指定要显示的模态框组件 id，对应下面代码中的 id="addCategory"。

```html
<div class="modal fade" id="addCategory" tabindex="-1">
    <div class="modal-dialog">
        <div class="modal-content">
            <div class="modal-header">
                <h5 class="modal-title">添加分类</h5>
                <button type="button" class="close" data-dismiss="modal">
                    <span>&times;</span>
                </button>
            </div>
            <div class="modal-body">
                <form id="addCategoryForm" action="/category/add" method="post">
                    <div class="form-group row">
                        <label class="col-sm-2 col-form-label">分类名</label>
                        <div class="col-sm-10">
                            <input type="text" name="name" class="form-
```

```html
control" placeholder="分类名">
                            </div>
                        </div>

                        <div class="form-group row">
                            <label class="col-sm-2 col-form-label">简介</label>
                            <div class="col-sm-10">
                                <textarea name="description" class="form-control" rows="4"></textarea>
                            </div>
                        </div>
                        <div class="text-danger d-none alert">
                        </div>
                    </form>
                </div>
                <div class="modal-footer">
                    <button type="button" class="btn btn-secondary" data-dismiss="modal">关闭</button>
                    <button id="addButton" type="button" class="btn btn-primary">提交</button>
                </div>
            </div>
        </div>
    </div>
```

代码 `<button id="addButton" type="button" class="btn btn-primary">` 中为"提交"按钮设置了 id 值,通过 jQuery 代码为该按钮绑定了 click 事件处理代码,当"提交"按钮被单击时,通过 CSS 选择器找到 id="addCategoryForm" 表单里面需要进行非空验证的控件进行数据验证。

`$("#addButton").click(function () { ... })` 用来为前文中 id="addButton"的按钮绑定 click 事件处理代码。事件处理代码中 `let addForm = $("#addCategoryForm");` 通过 jQuery 选择器返回 id="addCategoryForm" 的表单元素,通过该元素对象调用 find() 方法可以查找元素内部的指定子元素对象。例如,代码 addForm.find("input[name=name]") 中该元素对象通过属性选择器找到子元素中属性 name="name" 的 input 元素,通过 input 元素的 val()方法可以取到输入框中的值,如果没有输入,就在错误提示中进行提示。代码如下所示。

```
<script>
    $(document).ready(function () {
        $("#addButton").click(function () {
            let error = "";
            let addForm = $("#addCategoryForm");
            if (addForm.find("input[name=name]").val() == "") {
                error += "请输入分类名<br>";
            }

            if (error != "") {
                addForm.find("div[class~=alert]").html(error);
                addForm.find("div[class~=alert]").removeClass("d-none");
            } else {
                addForm.find("div[class~=alert]").addClass("d-none");
                addForm.submit();
            }
```

```
    });
});
</script>
```

8.3 删除分类

仔细阅读"删除分类"用例描述要求,理解时序图的含义,按要求完成编码工作。

8.3.1 任务需求

删除分类用例描述如表 8-3 所示。

表 8-3 删除分类用例描述

用例名称	删除分类
编号	07
参与者	管理员
简要说明	管理员删除指定分类
基本事件流	1. 管理员在查询表单中输入查询条件,单击"查询"按钮 2. 系统显示符合条件的分类 3. 管理员单击指定分类对应的"删除"超链接 4. 系统删除相应的分类记录 5. 系统返回分类查询页面
其他事件流	无
异常事件流	返回异常页面,提示错误信息
前置条件	管理员已经登录系统
后置条件	相应记录从数据库中删除

删除分类功能时序图如图 8-8 所示。

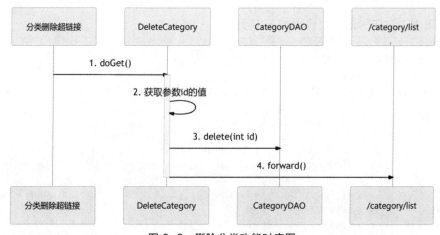

图 8-8 删除分类功能时序图

时序图中各参与者说明如下。

分类删除超链接：视图层，分类查询结果页面列表中的每一条分类记录都有一个对应的"删除"超链接，它会向后台传递要删除的分类主键。

DeleteCategory：控制层，位于 store.controller 包中，是一个 Servlet。其请求地址为/category/delete，处理来自分类删除超链接的 get 请求，从请求参数中获取需要删除的分类主键并调用 CategoryDAO 的 delete()方法删除数据，操作完成后返回分类查询视图。

CategoryDAO：数据访问模型层。其 delete()方法通过 JDBC 将 t_category 表中对应主键的记录删除。

/category/list：分类查询控制层对应的地址。请求该地址最终会返回分类查询视图层。

在分类查询结果列表中，每一条分类记录都有对应的"删除"超链接，单击"删除"超链接可以从数据库中删除与其对应的分类记录，如图 8-9 所示。

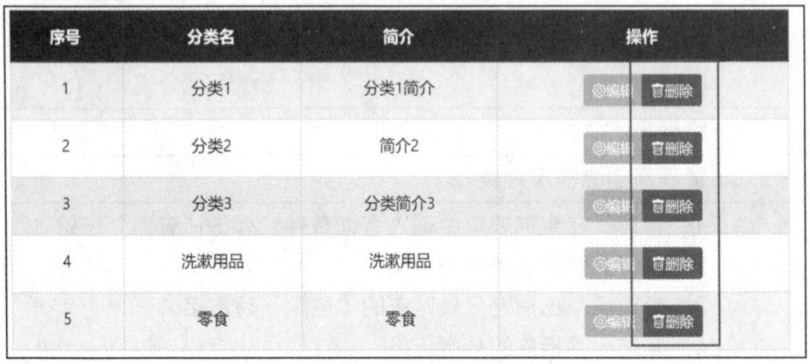

图 8-9　删除分类界面

8.3.2　任务实现

【任务 8-7】完成方法 CategoryDAO.delete(int id)

编辑 store.dao 下的 CategoryDAO 类，添加 delete(int id)方法。

CategoryDAO 用来封装对 t_category 表的操作，其中 delete(int id)方法负责根据传入的主键删除数据库中对应的记录，代码如下所示。

```java
package store.dao;
//省略包导入代码
public class CategoryDAO {
    public void delete(int id) throws Exception {
        Connection con= null;
        PreparedStatement ps=null;
        try {
            Class.forName(DB.JDBC_DRIVER);
            con=DriverManager.getConnection(DB.JDBC_URL,DB.JDBC_USER,DB.JDBC_PASSWORD);
            ps=con.prepareStatement("delete from t_category where id=?");
            ps.setInt(1,id);
            ps.executeUpdate();
        } catch (Exception e) {
```

```
                e.printStackTrace();
                throw new Exception("数据库异常:"+e.getMessage());
            }finally {
                if(ps!=null) ps.close();
                if(con!=null) con.close();
            }
        }
    }
```

【任务 8-8】完成 DeleteCategory.java

在 store.controller 包下新建 Servlet 类 DeleteCategory,其请求地址为/category/delete。

DeleteCategory 用来接收分类删除超链接的请求,从请求参数中获取需要删除的分类主键,然后将主键作为参数调用 CategoryDAO 的 delete(int id)方法。记录删除成功后转发到资源地址/category/list,该资源会进行一次无参数的分类查询。代码如下所示。

```
package store.controller;
//省略包导入代码
@WebServlet("/category/delete")
public class DeleteCategory extends HttpServlet {
    protected void doGet(HttpServletRequest request, HttpServletResponse response) throws ServletException, IOException {
        int id=Integer.parseInt(request.getParameter("id"));
        try {
            new CategoryDAO().delete(id);
            request.getRequestDispatcher("/category/list").forward(request,response);
        } catch (Exception e) {
            e.printStackTrace();
            throw new RuntimeException(e.getMessage());
        }
    }
}
```

【任务 8-9】完成视图部分

category.jsp 中包含分类查询结果显示代码,通过 JSTL 的 forEach 标签循环查询结果,每次循环生成一行结果,代码如下所示。

```
<c:forEach items="${requestScope.categoryList}" var="c" varStatus="vs">
    <tr>
        <td>${vs.count}</td>
        <td>${c.name}</td>

        <td class="text-truncate" style="max-width: 150px;">${c.description}</td>
        <td>
            <div class="btn-group">
                <a class="btn btn-sm btn-warning text-white" href="#"
                                                data-toggle="modal" data-target="#editCategory" category-id="${c.id}">
                    <i class="bi-gear"></i>编辑</a>
                <a class="btn btn-sm btn-danger" href="/category/delete?id=${c.id}">
                    <i class="bi-trash"></i>删除</a>
```

```
                    </div>
                </td>
            </tr>
</c:forEach>
```

其中生成"删除"超链接的代码 href="/category/delete?id=${c.id}" 通过 EL 表达式从当前循环的分类对象中取出主键并赋值给超链接的参数 id，当该超链接被单击时，其对应的主键就会被传递给后台。

8.4 修改分类

仔细阅读"修改分类"用例描述要求，理解时序图的含义，按要求完成编码工作。

8.4.1 任务需求

修改分类用例描述如表 8-4 所示。

表 8-4 修改分类用例描述

用例名称	修改分类
编号	08
参与者	管理员
简要说明	管理员修改指定分类
基本事件流	1. 管理员在分类查询结果页面单击需要修改的分类对应的"编辑"超链接 2. 系统显示分类编辑界面 3. 管理员在编辑界面编辑数据后单击"提交"按钮 4. 系统保存修改后的数据 5. 系统返回分类查询页面
其他事件流	无
异常事件流	返回异常页面，提示错误信息
前置条件	管理员已经登录系统
后置条件	分类数据被更新到数据库

修改分类功能时序图如图 8-10 所示。

时序图中各参与者说明如下。

category.jsp：视图层。分类查询结果页面列表中的每一条分类记录都有一个对应的"编辑"超链接，单击"编辑"超链接会发送一个 AJAX 请求，该请求将需要更新的分类主键传给后台。

GetCategory：控制层，位于 store.controller 包中，是一个 Servlet。其请求地址为/category，处理来自 category.jsp 的异步 get 请求，用于获取请求参数中的分类主键并调用 CategoryDAO

的 findById(int id)方法查询分类信息，操作完成后以 JSON 格式返回分类信息。

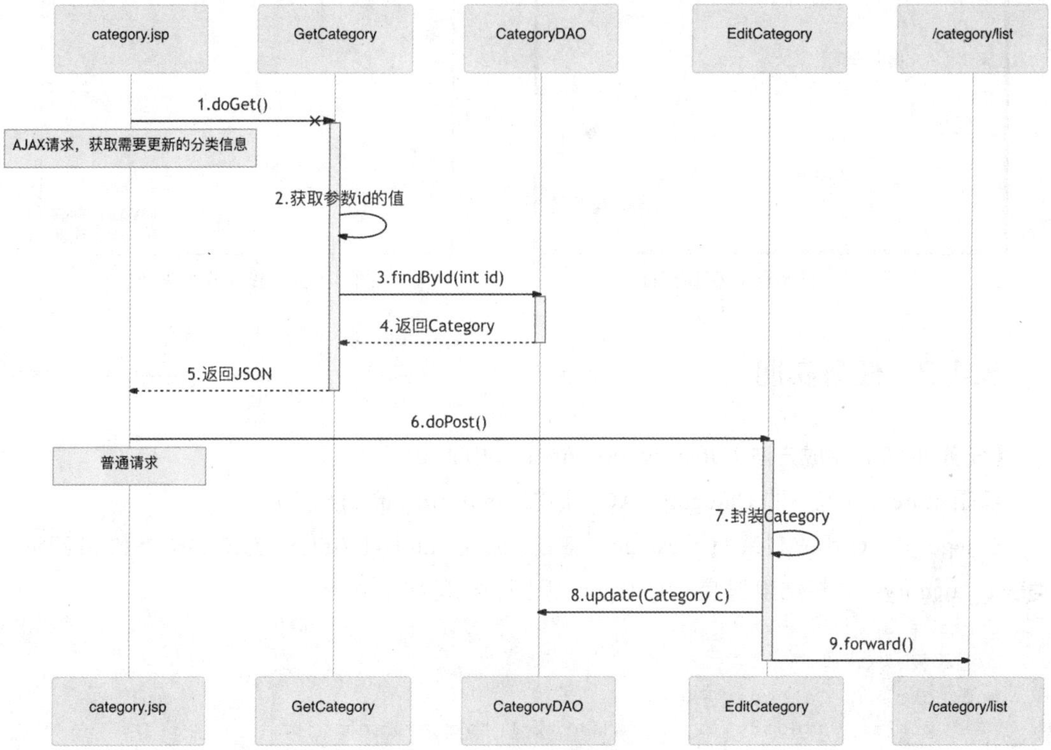

图 8-10 修改分类功能时序图

EditCategory：控制层，位于 store.controller 包中，是一个 Servlet。其请求地址为/category/edit，处理来自 category.jsp 的 post 请求，用于获取请求参数中的分类修改信息并将其封装成 Category 对象，调用 CategoryDAO 的 update(Category c) 方法更新分类信息，操作完成后返回分类查询页面。

CategoryDAO：数据访问模型层。其 findById(int id)方法根据 id 查找对应的分类信息并返回。

/category/list：分类查询控制层对应的地址，请求该地址最终会返回分类查询视图层。

【注意】

分类信息修改由两个步骤构成。第一步：通过 AJAX 请求查询需要修改的分类信息，服务器以 JSON 格式返回数据，前端 jQuery 代码接收到数据后填充分类信息编辑界面。第二步：管理员修改完成后单击"提交"按钮，通过表单以同步消息发送修改后的数据给服务器保存。

在分类查询结果页面单击分类对应的"编辑"超链接，弹出分类信息编辑界面，如图 8-11 所示。分类名是必填项，如果没有输入，在提交时会报错，如图 8-12 所示。修改成功后显示分类的查询结果页面。

图 8-11　分类信息编辑界面　　　　　　　图 8-12　修改分类失败

8.4.2　任务实现

【任务 8-10】完成方法 CategoryDAO.findById(int id)

编辑 store.dao 包下的 CategoryDAO，添加 findById(int id)方法。

CategoryDAO 用来封装对 t_category 表的操作。findById(int id)方法根据参数 id 查询分类表 t_category，并将结果封装成 Category 返回，代码如下所示。

```
package store.dao;
//省略包导入代码
public class CategoryDAO {
    public Category findById(int id) throws Exception {
        Category c=null;
        Connection con= null;
        PreparedStatement ps=null;
        ResultSet rs=null;
        try {
            Class.forName(DB.JDBC_DRIVER);
            con=DriverManager.getConnection(DB.JDBC_URL,DB.JDBC_USER,DB.JDBC_PASSWORD);
            ps=con.prepareStatement("select * from t_category where id=? ");
            ps.setInt(1,id);
            rs=ps.executeQuery();
            if (rs.next()){
                c=new Category();
                c.setId(rs.getInt("id"));
                c.setName(rs.getString("name"));
                c.setDescription(rs.getString("description"));
            }
        } catch (Exception e) {
            e.printStackTrace();
            throw new Exception("数据库异常:"+e.getMessage());
        }finally {
            if(rs!=null) rs.close();
            if(ps!=null) ps.close();
            if(con!=null) con.close();
        }
```

```
            return c;
        }
}
```

【任务 8-11】 完成方法 CategoryDAO.update(Category c)

编辑 store.dao 包下的 CategoryDAO 类，添加 update(Category c)方法。update(Category c)方法根据参数 c 的 id 值将其属性值更新到数据库对应的记录中，代码如下所示。

```
package store.dao;
//省略包导入代码
public class CategoryDAO {
    public void update(Category c) throws Exception {
        Connection con= null;
        PreparedStatement ps=null;
        try {
            Class.forName(DB.JDBC_DRIVER);
            con=DriverManager.getConnection(DB.JDBC_URL,DB.JDBC_USER,DB.JDBC_PASSWORD);
            ps=con.prepareStatement("update t_category set name=?,description=? where id=?");
            ps.setString(1,c.getName());
            ps.setString(2,c.getDescription());
            ps.setInt(3,c.getId());
            ps.executeUpdate();
        } catch (Exception e) {
            e.printStackTrace();
            throw new Exception("数据库异常:"+e.getMessage());
        }finally {
            if(ps!=null) ps.close();
            if(con!=null) con.close();
        }
    }
}
```

【任务 8-12】 完成 GetCategory.java

在 store.controller 包下新建 Servlet 类 GetCategory，其请求地址为/category。

GetCategory 的 doGet()方法接收前端的 AJAX 请求，通过请求获取参数分类主键，然后通过 CategoryDAO 查询对应的分类信息，最后以 JSON 格式返回数据。GetCategory 数据处理过程跟普通请求一样，重点在于返回数据的方式变化。此 Servlet 将数据转换成 JSON 对象，然后通过输出流将对象以 JSON 格式返回。代码如下所示。

```
package store.controller;
//省略包导入代码
@WebServlet("/category")
public class GetCategory extends HttpServlet {
    protected void doGet(HttpServletRequest request, HttpServletResponse response) throws ServletException, IOException {
        int id=Integer.parseInt(request.getParameter("id"));
        try {
            Category c=new CategoryDAO().findById(id);
            JSONObject jsonObject=new JSONObject();
            jsonObject.put("success",true);
            jsonObject.put("data",c);
```

```
                response.setContentType("application/x-json");
                response.getWriter().write(jsonObject.toJSONString());
                response.getWriter().flush();
                response.getWriter().close();
            } catch (Exception e) {
                e.printStackTrace();
                JSONObject jsonObject=new JSONObject();
                jsonObject.put("success",false);
                jsonObject.put("msg",e.getMessage());
                response.setContentType("application/x-json");
                response.getWriter().write(jsonObject.toJSONString());
                response.getWriter().flush();
                response.getWriter().close();
            }
        }
    }
```

【任务 8-13】完成 EditCategory.java

在 store.controller 包下新建 Servlet 类 EditCategory，其请求地址为/category/edit。

EditCategory 接收分类更新表单的 post 请求，从请求参数中获取更新数据并将其封装成 Category 对象，然后调用 CategoryDAO 的 update(Category c)方法更新数据，完成后转发到资源路径/category/list，该资源路径会返回分类查询结果页面。代码如下所示。

```
package store.controller;
//省略包导入代码
@WebServlet("/category/edit")
public class EditCategory extends HttpServlet {
    protected void doPost(HttpServletRequest request, HttpServletResponse response) throws ServletException, IOException {
        Category c=new Category();
        c.setId(Integer.parseInt(request.getParameter("id")));
        c.setName(request.getParameter("name"));
        c.setDescription(request.getParameter("description"));
        try {
            new CategoryDAO().update(c);
            request.getRequestDispatcher("/category/list").forward(request, response);
        } catch (Exception e) {
            e.printStackTrace();
            throw new RuntimeException(e.getMessage());
        }
    }
}
```

【任务 8-14】完成视图部分

分类信息编辑界面通过"编辑"超链接触发编辑模态框。表单提交前需要进行数据验证，验证通过后通过 jQuery 代码提交，因此"提交"按钮的类型是 button，而不是 submit，代码如下所示。

```
<!-- 编辑分类 Modal -->
<div class="modal fade" id="editCategory">
    <div class="modal-dialog">
        <div class="modal-content">
```

```html
                <div class="modal-header">
                    <h5 class="modal-title">编辑分类</h5>
                    <button type="button" class="close" data-dismiss="modal">
                        <span>&times;</span>
                    </button>
                </div>
                <div class="modal-body">
                    <form id="editCategoryForm"action="/category/edit"method="post">
                        <div class="form-group row">
                            <label class="col-sm-2 col-form-label">分类</label>
                            <div class="col-sm-10">
                                <input type="hidden" name="id">
                                <input type="text" name="name" class="form-control" placeholder="分类名">
                            </div>
                        </div>
                        <div class="form-group row">
                            <label class="col-sm-2 col-form-label">简介</label>
                            <div class="col-sm-10">
                                <textarea name="description" class="form-control" rows="4"></textarea>
                            </div>
                        </div>
                        <div class="text-danger d-none alert">
                        </div>
                    </form>
                </div>
                <div class="modal-footer">
                    <button type="button" class="btn btn-secondary" data-dismiss="modal">关闭</button>
                    <button id="editButton" type="button" class="btn btn-primary">提交</button>
                </div>
            </div>
        </div>
    </div>
```

在实现分类查询结果页面时循环输出表格的行数据，每一行都有对应的"编辑"超链接，每个超链接都含有一个自定义属性 category-id，该属性的值为对应的分类主键。"编辑"超链接代码如下。

```html
<a class="btn btn-sm btn-warning text-white" href="#"  category-id="${c.id}"
    data-toggle="modal" data-target="#editCategory" ><i class="bi-gear"></i>
编辑</a>
```

通过 jQuery 代码 $("a[category-id]").click() 可以选中页面中所有的"编辑"超链接，并且为它们绑定单击事件。在单击事件处理代码中通过$.getJson()向资源/category 发送 AJAX 请求，在接收到返回值后使用返回的 JSON 数据为模态框中的控件赋值。代码如下所示。

```html
<script>
    $(document).ready(function () {
        $("a[category-id]").click(function () {
            $.getJSON("/category", {id: $(this).attr("category-id")}, function
```

```
(result) {
                    $("#editCategoryForm").find("input[name=id]").val(result.data.id);
                    $("#editCategoryForm").find("input[name=name]").val
(result.data.name);
                    $("#editCategoryForm").find("textarea[name=description]")
                            .val(result.data.description);
                });
            });
        });
    </script>
```

修改完成后单击"提交"按钮，需要对输入的数据进行数据校验。通过 jQuery 代码 $("#editButton").click()为"提交"按钮绑定单击事件，在事件处理代码中依次对必填数据进行校验，如果验证失败，通过 jQuery 找到表单中的警告信息显示层，并在其中显示信息。验证通过则提交表单。代码如下所示。

```
    <script>
        $(document).ready(function () {
            $("#editButton").click(function () {
                let error = "";
                let editForm = $("#editCategoryForm");
                if (editForm.find("input[name=name]").val() == "") {
                    error += "请输入分类名<br>";
                }

                if (error != "") {
                    editForm.find("div[class~=alert]").html(error);
                    editForm.find("div[class~=alert]").removeClass("d-none");
                } else {
                    editForm.find("div[class~=alert]").addClass("d-none");
                    editForm.submit();
                }
            });
        });
    </script>
```

本章习题

1. 结合项目代码，实现查询分类功能。
2. 结合项目代码，实现添加分类功能。
3. 结合项目代码，实现删除分类功能。
4. 结合项目代码，实现修改分类功能。

第9章
商品管理模块实现

本章目标

- 理解商品管理需求
- 理解商品数据表设计
- 理解商品管理各功能时序图
- 掌握 Java 开发技巧
- 掌握 Bootstrap 开发技巧
- 掌握 jQuery 开发技巧

通过前一章的学习，读者已经实现了分类管理模块的功能。本章通过实现超市管理系统的商品管理模块功能，帮助读者加深对 Bootstrap、jQuery 及 Servlet 等知识点的理解，锻炼使用编程语言解决实际问题的能力，强化基于 MVC 模式的编程思维。

9.1 查询商品

仔细阅读"查询商品"用例描述要求，理解时序图的含义，按要求完成编码工作。

9.1.1 任务需求

查询商品用例描述如表 9-1 所示。

表 9-1　查询商品用例描述

用例名称	查询商品
编号	09
参与者	管理员
简要说明	管理员输入条件查询商品
基本事件流	1. 管理员在查询表单中输入查询条件，单击"查询"按钮 2. 系统显示符合条件的商品
其他事件流	无
异常事件流	返回异常页面，提示错误信息
前置条件	管理员已经登录系统
后置条件	无

查询商品功能时序图如图 9-1 所示。

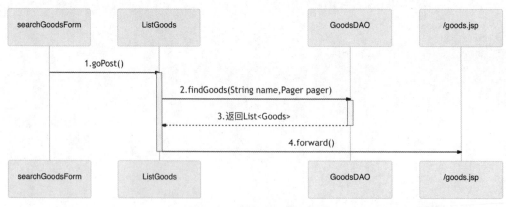

图 9-1　查询商品功能时序图

时序图中各参与者说明如下。

searchGoodsForm：视图层，放置于 goods.jsp 里面的表单，提交方式为 post，提交地址为/goods/list，包含控件分类名 name。

ListGoods：控制层，位于 store.controller 包中，是一个 Servlet。其请求地址为/goods/list，处理来自 searchGoodsForm 的 post 请求，用于获取查询条件分类名 name 并调用 GoodsDAO 的 findGoods(String name，Pager pager)方法进行分页查询，操作完成后返回商品查询视图。当单击"上一页"或者"下一页"按钮进行数据查询时，超链接发送的是 get 方式的请求，由于其处理逻辑和表单 post 方式请求完全相同，因此 ListGoods 中 doGet()方法直接调用 doPost()方法。

GoodsDAO：数据访问模型层。其 findGoods(String name，Pager pager)方法进行分页查询，findGoods(String name)方法进行不需要分页的查询。

/goods.jsp：商品管理页面的地址。请求该地址最终会返回商品管理视图层。

在商品管理主页面的条件输入框中输入查询条件并单击"查询"按钮，分页显示满足条件的查询结果，如图 9-2 所示。

图 9-2 查询商品

9.1.2 任务实现

【任务 9-1】完成方法 GoodsDAO.findGoods(String name, Pager pager)

在 store.dao 包下新建 GoodsDAO 类，并在其中添加方法 findGoods(String name, Pager pager)。

GoodsDAO 用来封装对 t_goods 表的操作，其 findGoods(String name, Pager pager)方法用于进行分页查询，其中 name 为查询条件商品名，pager 为分页帮助对象，代码如下所示。

```java
public List<Goods> findGoods(String name) throws Exception {
    List<Goods> list=new ArrayList<>();
    Connection con= null;
    PreparedStatement ps=null;
    ResultSet rs=null;
    try {
        Class.forName(DB.JDBC_DRIVER);
        con=DriverManager.getConnection(DB.JDBC_URL,DB.JDBC_USER,DB.JDBC_PASSWORD);
        ps=con.prepareStatement("select * from t_goods where name like ? ");
        ps.setString(1,"%"+name+"%");
        rs=ps.executeQuery();
        while (rs.next()){
            Goods g=new Goods();
            g.setId(rs.getInt("id"));
            g.setName(rs.getString("name"));
            g.setSpecs(rs.getString("specs"));
            g.setSn(rs.getString("sn"));
```

```
                g.setPrice(rs.getDouble("price"));
                g.setStock(rs.getInt("stock"));
                g.setCategoryId(rs.getInt("category_id"));
                list.add(g);
            }
        } catch (Exception e) {
            e.printStackTrace();
            throw new Exception("数据库异常:"+e.getMessage());
        }finally {
            if(rs!=null) rs.close();
            if(ps!=null) ps.close();
            if(con!=null) con.close();
        }
        return list;
    }
```

【任务 9-2】完成 ListGoods.java

在 store.controller 包下新建 Servlet 类 ListGoods，其请求地址为/goods/list。

ListGoods 的 doPost()方法用来处理商品查询表单的请求，doGet()方法用来处理分页按钮的请求，由于两者的逻辑相同，因此在 doGet()方法中直接调用 doPost()方法即可。ListGoods 从请求参数中得到查询条件 name，从请求对象的属性中得到分页对象 pager，将两者作为参数对 GoodsDAO 的 findGoods()方法进行调用，得到分页后的查询结果 goodsList。查询完毕后通过 request 对象的属性将 pager 和 goodsList 两个对象转发到 goods.jsp 页面，代码如下所示。

```
@WebServlet("/goods/list")
public class ListGoods extends HttpServlet {
    protected void doPost(HttpServletRequest request, HttpServletResponse response) throws ServletException, IOException {
        String name=request.getParameter("name");
        if(name==null) name="";
        try {
            Pager pager = (Pager) request.getAttribute("pager");
            List<Goods> goodsList=new GoodsDAO().findGoods(name,pager);
            request.setAttribute("goodsList",goodsList);
            request.getRequestDispatcher("/goods.jsp").forward(request,response);
        } catch (Exception e) {
            e.printStackTrace();
            throw new RuntimeException(e.getMessage());
        }
    }

    protected void doGet(HttpServletRequest request, HttpServletResponse response) throws ServletException, IOException {
        doPost(request,response);
    }
}
```

【任务 9-3】完成视图部分

查询商品表单的实现代码 searchGoodsForm 位于 goods.jsp 中，以 post 方式提交到/goods/list，提交参数 name 表示要查询的商品名，代码如下所示。

```
<form class="form-inline" action="/goods/list" method="post">
    <input type="text" name="name" class="form-control" placeholder="请输入
```

查询条件">
```
        <button type="submit" class="btn btn-success ml-2"><i class="bi-search"></i>查询</button>
    </form>
```

ListGoods 会通过 request 作用域返回两个对象，分别是分页查询结果集 goodsList 和分页对象 pager，在 goods.jsp 中通过循环 goodsList 生成结果表格，代码如下所示。

```
<table class="table table-hover table-bordered table-striped text-center">
    <thead class="thead-dark">
    <tr>
        <th scope="col">序号</th>
        <th scope="col">商品名称</th>
        <th scope="col">规格</th>
        <th scope="col">条形码</th>
        <th scope="col">售价</th>
        <th scope="col">库存</th>
        <th scope="col">操作</th>
    </tr>
    </thead>
    <tbody>
    <c:forEach items="${requestScope.goodsList}" var="g" varStatus="vs">
        <tr>
            <td>${vs.count}</td>
            <td>${g.name}</td>
            <td>${g.specs}</td>
            <td>${g.sn}</td>
            <td>${g.price}</td>
            <td>${g.stock}</td>
            <td>
                <div class="btn-group">
                    <a class="btn btn-sm btn-primary text-white" href="#"
                       data-toggle="modal" data-target="#addRestock" goods-id="${g.id}" goods-name="${g.name}">
                        <i class="bi-gear"></i>进货</a>
                    <a class="btn btn-sm btn-warning text-white" href="#"
                       data-toggle="modal" data-target="#editGoods" goods-id="${g.id}">
                        <i class="bi-gear"></i>编辑</a>
                    <a class="btn btn-sm btn-danger" href="/goods/delete?id=${g.id}">
                        <i class="bi-trash"></i>删除</a>
                </div>
            </td>
        </tr>
    </c:forEach>
    </tbody>
</table>
```

分页按钮部分代码如下所示，其中超链接请求/goods/list 进行分页查询；传递参数 currentPage 表示希望跳转到第几页，从分页对象 pager 中获取；name 为输入的查询条件，从请求参数中获取。

```
    <nav>
        <ul class="pagination justify-content-between">
            <li class="page-item">
                <a class="page-link rounded-pill"
href="/goods/list?currentPage=${requestScope.pager.previous}&name=${param.name}"
                   tabindex="-1">上一页</a>
            </li>
            <li class="page-item"><a class="page-link rounded-pill">总共${requestScope.pager.pageCount}页,当前第${requestScope.pager.currentPage}页</a>
            </li>
            <li class="page-item">
                <a class="page-link rounded-pill"
href="/goods/list?currentPage=${requestScope.pager.next}&name=${param.name}">
下一页</a>
            </li>
        </ul>
    </nav>
```

9.2 添加商品

仔细阅读"添加商品"用例描述要求,理解时序图的含义,按要求完成编码工作。

9.2.1 任务需求

添加商品用例描述如表 9-2 所示。

表 9-2 添加商品用例描述

用例名称	添加商品
编号	10
参与者	管理员
简要说明	管理员通过商品添加页面添加新的商品
基本事件流	1. 管理员单击"添加商品"按钮 2. 系统显示添加商品表单 3. 管理员在表单中输入商品名、规格、条形码、售价、分类 4. 管理员单击"提交"按钮,新商品被保存
其他事件流	在单击"提交"按钮时,如果有必填内容未填写,则提示管理员,数据不会被保存
异常事件流	返回异常页面,提示错误信息
前置条件	管理员已经登录系统
后置条件	商品被保存到数据库

添加商品功能时序图如图 9-3 所示。

第 9 章 商品管理模块实现

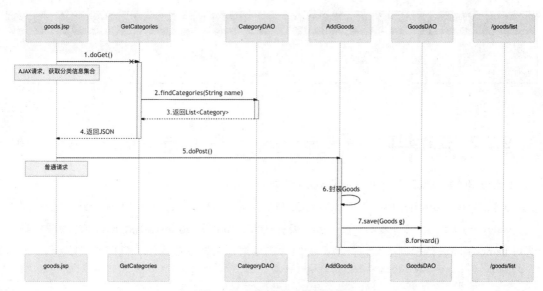

图 9-3　添加商品功能时序图

时序图中各参与者说明如下。

goods.jsp：视图层。单击"添加商品"按钮会发送一个 AJAX 请求，请求后台查询所有的商品分类信息，显示添加商品表单，并根据返回的 JSON 数据生成商品分类列表的内容。

GetCategories：控制层，位于 store.controller 包中，是一个 Servlet。其请求地址为 /categories，处理来自 goods.jsp 的异步 get 请求，调用 CategoryDAO 的 findCategories(String name) 方法查询分类信息，操作完成后以 JSON 格式返回分类信息。

AddGoods：控制层，位于 store.controller 包中，是一个 Servlet。其请求地址为 /goods/add，处理来自 goods.jsp 的 post 请求，用于获取请求参数中的商品信息并将其封装成 Goods 对象，调用 GoodsDAO 的 save(Goods g) 方法添加商品信息，操作完成后返回商品查询页面。

CategoryDAO：数据访问模型层。其 findCategories(String name) 方法根据分类名模糊查找 t_category 表，将符合条件的分类信息返回。

GoodsDAO：数据访问模型层。其 save(Goods g) 方法将一个 Goods 对象的值通过 JDBC 添加到 t_goods 表中。

/goods/list：商品查询控制层对应的地址，请求该地址最终会返回商品查询视图层。

在商品管理页面单击"添加商品"按钮，弹出商品添加界面。添加商品时商品名、规格、条形码、售价、分类是必填项，如果没有输入，在提交时会报错，如图 9-4 所示。添加成功后显示商品的查询结果页面，新添加的商品库存数量默认为 0，在后文介绍的进货管理功能中可以对其进行管理，如图 9-5 所示。

图 9-4　添加商品校验失败

187

图 9-5　添加商品成功

9.2.2　任务实现

【任务 9-4】完成方法 CategoryDAO.findCategories(String name)

编辑 store.controller 包下的 CategoryDAO 类，添加 findCategories(String name)方法。

CategoryDAO 用来封装对 t_category 表的操作，findCategories(String name)方法根据参数值对 name 字段进行模糊查询，并将符合条件的结果封装成集合返回，代码如下所示。

```java
public List<Category> findCategories(String name) throws Exception {
    List<Category> list=new ArrayList<>();
    Connection con= null;
    PreparedStatement ps=null;
    ResultSet rs=null;
    try {
        Class.forName(DB.JDBC_DRIVER);
        con=DriverManager.getConnection(DB.JDBC_URL,DB.JDBC_USER,DB.JDBC_PASSWORD);
        ps=con.prepareStatement("select * from t_category where name like ? ");
        ps.setString(1,"%"+name+"%");
        rs=ps.executeQuery();
        while (rs.next()){
            Category c=new Category();
            c.setId(rs.getInt("id"));
            c.setName(rs.getString("name"));
            c.setDescription(rs.getString("description"));
            list.add(c);
        }
    } catch (Exception e) {
        e.printStackTrace();
        throw new Exception("数据库异常:"+e.getMessage());
    }finally {
        if(rs!=null) rs.close();
        if(ps!=null) ps.close();
        if(con!=null) con.close();
    }
    return list;
}
```

【任务 9-5】完成 GetCategories.java

在 store.controller 包下新建 Servlet 类 GetCategories，其请求地址为/categories。

GetCategories 的 doGet()方法接收前端的 AJAX 请求，然后通过 CategoryDAO 查询分类信息集合，最后以 JSON 格式返回数据，代码如下所示。

```java
@WebServlet("/categories")
public class GetCategories extends HttpServlet {
```

```java
        protected void doGet(HttpServletRequest request, HttpServletResponse
response) throws ServletException, IOException {
            String name=request.getParameter("name");
            if(name==null) name="";
            try {
                List<Category> list=new CategoryDAO().findCategories(name);
                JSONObject jsonObject=new JSONObject();
                jsonObject.put("success",true);
                jsonObject.put("data",list);
                response.setContentType("application/x-json");
                response.getWriter().write(jsonObject.toJSONString());
                response.getWriter().flush();
                response.getWriter().close();
            } catch (Exception e) {
                e.printStackTrace();
                JSONObject jsonObject=new JSONObject();
                jsonObject.put("success",false);
                jsonObject.put("msg",e.getMessage());
                response.setContentType("application/x-json");
                response.getWriter().write(jsonObject.toJSONString());
                response.getWriter().flush();
                response.getWriter().close();
            }
        }
    }
```

【任务 9-6】完成方法 GoodsDAO.save(Goods g)

编辑 store.controller 包下的 GoodsDAO 类，添加 save(Goods g)方法。

GoodsDAO 用来封装对 t_goods 表的操作，save(Goods g)方法将 Goods 类型参数 g 的属性值保存成 t_goods 表中的一条记录，代码如下所示。

```java
    public void save(Goods g) throws Exception {
        Connection con= null;
        PreparedStatement ps=null;
        try {
            Class.forName(DB.JDBC_DRIVER);
            con=DriverManager.getConnection(DB.JDBC_URL,DB.JDBC_USER,DB.JDBC_PASSWORD);
            ps=con.prepareStatement("insert into t_goods value (null ,?,?,?,?,?,?)");
            ps.setString(1,g.getName());
            ps.setString(2,g.getSpecs());
            ps.setString(3,g.getSn());
            ps.setDouble(4,g.getPrice());
            ps.setInt(5,g.getStock());
            ps.setInt(6,g.getCategoryId());
            ps.executeUpdate();
        } catch (Exception e) {
            e.printStackTrace();
            throw new Exception("数据库异常:"+e.getMessage());
        }finally {
            if(ps!=null) ps.close();
            if(con!=null) con.close();
        }
    }
```

【任务 9-7】完成 AddGoods.java

在 store.controller 包下新建 Servlet 类 AddGoods，其请求地址为/goods/add。

AddGoods 用来接收 addGoodsForm 表单的请求数据。它将数据封装成一个 Goods 对象，然后调用 GoodsDAO 的 save()方法进行保存。需要注意的是，因为是添加操作，被添加的数据还没有主键，所以在封装被添加对象时不需要为 id 赋值。保存成功后转发到资源地址/goods/list，代码如下所示。

```java
@WebServlet("/goods/add")
public class AddGoods extends HttpServlet {
    protected void doPost(HttpServletRequest request, HttpServletResponse response) throws ServletException, IOException {
        Goods g=new Goods();
        g.setName(request.getParameter("name"));
        g.setSpecs(request.getParameter("specs"));
        g.setSn(request.getParameter("sn"));
        g.setPrice(Double.parseDouble(request.getParameter("price")));
        g.setStock(0);
        g.setCategoryId(Integer.parseInt(request.getParameter("category_id")));

        try {
            new GoodsDAO().save(g);
            request.getRequestDispatcher("/goods/list").forward(request,response);
        } catch (Exception e) {
            e.printStackTrace();
            throw new RuntimeException(e.getMessage());
        }
    }
}
```

【任务 9-8】完成视图部分

添加商品信息输入界面通过 goods.jsp 页面中的"添加商品"按钮触发模态框生成，代码如下所示。

```html
<button type="button" class="btn btn-primary" data-toggle="modal" data-target="#addGoods">
    <i class="bi-plus"></i>添加商品
</button>
```

代码中 data-toggle="modal"用来指定 Bootstrap 的触发行为是模态框，data-target="#addGoods" 用来指定要显示的模态框组件 id，对应下面代码中的 id="addGoods"。

```html
<div class="modal fade" id="addGoods" tabindex="-1">
    <div class="modal-dialog">
        <div class="modal-content">
            <div class="modal-header">
                <h5 class="modal-title">添加商品</h5>
                <button type="button" class="close" data-dismiss="modal">
                    <span>&times;</span>
                </button>
            </div>
            <div class="modal-body">
                <form id="addGoodsForm" action="/goods/add" method="post">
                    <div class="form-group row">
```

```html
                                <label class="col-sm-2 col-form-label">商品</label>
                                <div class="col-sm-10">
                                    <input type="text" name="name" class="form-control" placeholder="商品名">
                                </div>
                            </div>
                            <div class="form-group row">
                                <label class="col-sm-2 col-form-label">规格</label>
                                <div class="col-sm-10">
                                    <input type="text" name="specs" class="form-control" placeholder="规格">
                                </div>
                            </div>

                            <div class="form-group row">
                                <label class="col-sm-2 col-form-label">条形码</label>
                                <div class="col-sm-10">
                                    <input type="text" name="sn" class="form-control" placeholder="条形码">
                                </div>
                            </div>
                            <div class="form-group row">
                                <label class="col-sm-2 col-form-label">售价</label>
                                <div class="col-sm-10">
                                    <div class="input-group mb-3">
                                        <input type="number" name="price" class="form-control" placeholder="售价">
                                        <div class="input-group-append">
                                            <span class="input-group-text">元</span>
                                        </div>
                                    </div>
                                </div>
                            </div>
                            <div class="form-group row">
                                <label class="col-sm-2 col-form-label">分类</label>
                                <div class="col-sm-10">
                                    <select name="category_id"class="form-control">
                                        <option>请选择</option>
                                    </select>
                                </div>
                            </div>
                            <div class="text-danger d-none alert">
                            </div>
                        </form>
                    </div>
                    <div class="modal-footer">
                        <button type="button" class="btn btn-secondary" data-dismiss="modal">关闭</button>
                        <button id="addButton" type="button" class="btn btn-primary">提交</button>
                    </div>
```

```
            </div>
        </div>
    </div>
```

添加商品表单中的"分类"下拉列表默认只有一个"请选择"选项,当管理员单击"添加商品"按钮时,需要通过 AJAX 请求资源地址/categories 查询分类集合,并根据返回结果生成下拉列表项,代码如下。

```
$("button[data-target='#addGoods']").click(function () {
    $.getJSON("/categories", function (result) {
        let list = result.data;
        $("#addGoodsForm select[name=category_id]").empty();
        $("#addGoodsForm select[name=category_id]").append('<option value="">请选择</option>');
        list.forEach(({id, name}) => {
            $("#addGoodsForm select[name=category_id]").append('<option value="\${id}">\${name}</option>');//避免$和 EL 表达式冲突,使用转义字符
        })
    })
});
```

代码 `<button id="addButton" type="button" class="btn btn-primary">` 中为"提交"按钮设置了 id 值,通过 jQuery 代码为该按钮绑定了 click 事件处理代码,当"提交"按钮被单击时,通过 CSS 选择器找到 id="addGoodsForm" 表单里面需要进行非空验证的控件进行数据验证。代码如下所示。

```
$("#addButton").click(function () {
    let error = "";
    let addForm = $("#addGoodsForm");
    if (addForm.find("input[name=name]").val() == "") {
        error += "请输入商品名<br>";
    }
    if (addForm.find("input[name=specs]").val() == "") {
        error += "请输入商品规格<br>";
    }
    if (addForm.find("input[name=sn]").val() == "") {
        error += "请输入商品条形码<br>";
    }
    if (addForm.find("input[name=price]").val() == "") {
        error += "请输入商品价格<br>";
    }
    if (addForm.find("select[name=category_id]").val() == "") {
        error += "请选择分类<br>";
    }
    if (error != "") {
        addForm.find("div[class~=alert]").html(error);
        addForm.find("div[class~=alert]").removeClass("d-none");
    } else {
        addForm.find("div[class~=alert]").addClass("d-none");
        addForm.submit();
    }
});
```

9.3 删除商品

仔细阅读"删除商品"用例描述要求,理解时序图的含义,按要求完成编码工作。

9.3.1 任务需求

删除商品用例描述如表 9-3 所示。

表 9-3 删除商品用例描述

用例名称	删除商品
编号	11
参与者	管理员
简要说明	管理员删除指定商品
基本事件流	1. 管理员在查询表单中输入查询条件,单击"查询"按钮 2. 系统显示符合条件的商品 3. 管理员单击指定商品对应的"删除"超链接 4. 系统删除相应的商品记录 5. 系统返回商品查询页面
其他事件流	无
异常事件流	返回异常页面,提示错误信息
前置条件	管理员已经登录系统
后置条件	相应记录从数据库中删除

删除商品功能时序图如图 9-6 所示。

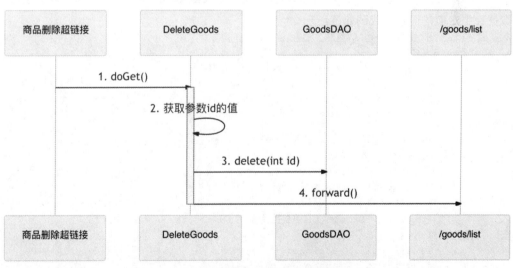

图 9-6 删除商品功能时序图

时序图中各参与者说明如下。

商品删除超链接：视图层，商品查询结果页面列表中的每一条商品记录都有一个对应的"删除"超链接，商品删除超链接会向后台传递要删除的商品主键。

DeleteGoods：控制层，位于 store.controller 包中，是一个 Servlet。其请求地址为/goods/delete，处理来自商品删除超链接的 get 请求，从请求参数中获取需要删除的商品主键并调用 GoodsDAO 的 delete()方法删除数据，操作完成后返回商品查询视图。

GoodsDAO：数据访问模型层。其 delete()方法通过 JDBC 将 t_goods 表中对应主键的记录删除。

/goods/list：商品查询控制层对应的地址。请求该地址最终会返回商品查询视图层。

在商品查询结果列表中每一条商品记录都有对应的"删除"超链接，单击"删除"超链接可以从数据库中删除其对应的商品记录，如图 9-7 所示。

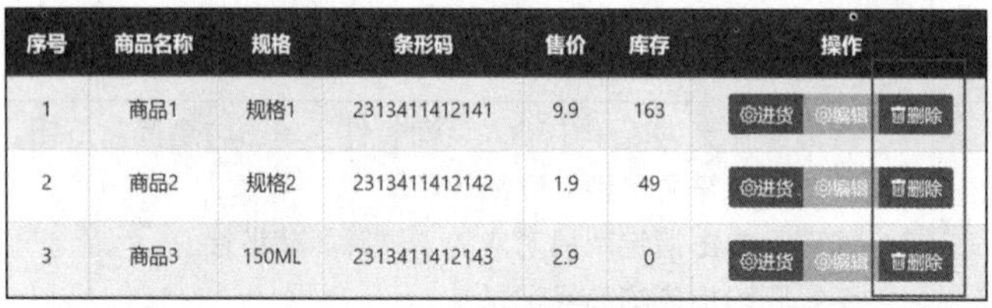

序号	商品名称	规格	条形码	售价	库存	操作
1	商品1	规格1	2313411412141	9.9	163	◎进货 ◎编辑 ⬛删除
2	商品2	规格2	2313411412142	1.9	49	◎进货 ◎编辑 ⬛删除
3	商品3	150ML	2313411412143	2.9	0	◎进货 ◎编辑 ⬛删除

图 9-7　删除商品界面

9.3.2　任务实现

【任务 9-9】完成方法 GoodsDAO.delete(int id)

编辑 store.dao 下的 GoodsDAO 类，添加 delete(int id)方法。

GoodsDAO 用来封装对 t_goods 表的操作，其中 delete(int id)方法负责根据传入的主键删除数据库中对应的记录，代码如下所示。

```java
public void delete(int id) throws Exception {
    Connection con= null;
    PreparedStatement ps=null;
    try {
        Class.forName(DB.JDBC_DRIVER);
        con=DriverManager.getConnection(DB.JDBC_URL,DB.JDBC_USER,DB.JDBC_PASSWORD);
        ps=con.prepareStatement("delete from t_goods where id=?");
        ps.setInt(1,id);
        ps.executeUpdate();
    } catch (Exception e) {
        e.printStackTrace();
        throw new Exception("数据库异常:"+e.getMessage());
    }finally {
        if(ps!=null) ps.close();
        if(con!=null) con.close();
    }
}
```

【任务 9-10】完成 DeleteGoods.java

在 store.controller 包下新建 Servlet 类 DeleteGoods，其请求地址为/goods/delete。

DeleteGoods 用来接收商品删除超链接的请求，并从请求参数中获取需要删除的商品主键，然后将主键作为参数调用 GoodsDAO 的 delete(int id)方法。记录删除成功后转发到资源地址/goods/list，该资源会进行一次无参数的商品查询。代码如下所示。

```
@WebServlet("/goods/delete")
public class DeleteGoods extends HttpServlet {

    protected void doGet(HttpServletRequest request, HttpServletResponse response) throws ServletException, IOException {
        int id=Integer.parseInt(request.getParameter("id"));
        try {
            new GoodsDAO().delete(id);
            request.getRequestDispatcher("/goods/list").forward(request, response);
        } catch (Exception e) {
            e.printStackTrace();
            throw new RuntimeException(e.getMessage());
        }
    }
}
```

【任务 9-11】完成视图部分

goods.jsp 中包含分类查询结果显示代码，通过 JSTL 的 forEach 标签循环查询结果，每次循环生成一行结果，代码如下所示。

```
<c:forEach items="${requestScope.goodsList}" var="g" varStatus="vs">
    <tr>
        <td>${vs.count}</td>
        <td>${g.name}</td>
        <td>${g.specs}</td>
        <td>${g.sn}</td>
        <td>${g.price}</td>
        <td>${g.stock}</td>
        <td>
            <div class="btn-group">
                <a class="btn btn-sm btn-primary text-white" href="#"
                    data-toggle="modal" data-target="#addRestock" goods-id="${g.id}" goods-name="${g.name}">
                    <i class="bi-gear"></i>进货</a>
                <a class="btn btn-sm btn-warning text-white" href="#"
                    data-toggle="modal" data-target="#editGoods" goods-id="${g.id}">
                    <i class="bi-gear"></i>编辑</a>
                <a class="btn btn-sm btn-danger" href="/goods/delete?id=${g.id}">
                    <i class="bi-trash"></i>删除</a>
            </div>
        </td>
    </tr>
</c:forEach>
```

其中生成"删除"超链接的代码 href="/goods/delete?id=${g.id}" 通过 EL 表达式从当前循环的商品对象中取出主键并赋值给超链接的参数 id，当该超链接被单击时，其对应的主键就会被传递给后台。

9.4 修改商品

仔细阅读"修改商品"用例描述要求，理解时序图的含义，按要求完成编码工作。

9.4.1 任务需求

修改商品用例描述如表 9-4 所示。

表 9-4 修改商品用例描述

用例名称	修改商品
编号	12
参与者	管理员
简要说明	管理员修改指定商品
基本事件流	1. 管理员在商品查询结果页面单击需要更新的商品对应的"编辑"超链接 2. 系统显示商品编辑界面 3. 管理员在编辑界面编辑数据后单击"提交"按钮 4. 系统保存修改后的数据 5. 系统返回商品查询页面
其他事件流	无
异常事件流	返回异常页面，提示错误信息
前置条件	管理员已经登录系统
后置条件	商品数据被更新到数据库

修改商品功能时序图如图 9-8 所示。

时序图中各参与者说明如下。

goods.jsp：视图层。商品查询结果页面列表中的每一条商品记录都有一个对应的"编辑"超链接，单击"编辑"超链接会发送两次异步请求，分别请求商品分类集合和需要修改的商品信息。

GetCategories：控制层，位于 store.controller 包中，是一个 Servlet。其请求地址为/categories，处理来自 goods.jsp 的异步 get 请求，调用 CategoryDAO 的 findCategories(String name)方法查询分类信息，操作完成后以 JSON 格式返回分类信息。

GetGoods：控制层，位于 store.controller 包中，是一个 Servlet。其请求地址为/goods，

处理来自 goods.jsp 的异步 get 请求,用于获取请求参数中的商品主键并调用 GoodsDAO 的 findById(int id)方法查询商品信息,操作完成后以 JSON 格式返回分类信息。

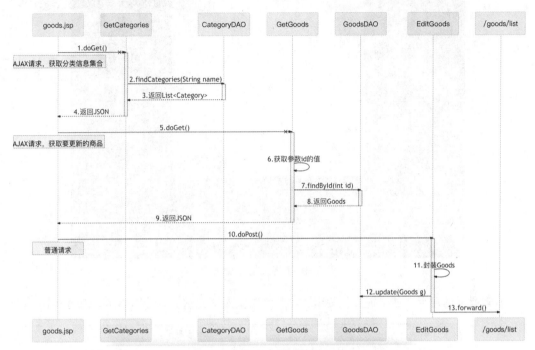

图 9-8 修改商品功能时序图

EditGoods:控制层,位于 store.controller 包中,是一个 Servlet。其请求地址为/goods/edit,处理来自 goods.jsp 的 post 请求,用于获取请求参数中的商品修改信息并将其封装成 Goods 对象,调用 GoodsDAO 的 update(Goods g) 方法更新商品信息,操作完成后返回商品查询页面。

CategoryDAO:数据访问模型层。其 findCategories(String name)方法根据分类名模糊查找 t_category 表,将符合条件的分类信息返回。

GoodsDAO:数据访问模型层。其 findById(int id)方法根据 id 查找对应的商品信息并返回。

/goods/list:商品查询控制层对应的地址,请求该地址最终会返回商品查询视图层。

【注意】

商品信息修改由 3 个步骤构成。第一步:通过 AJAX 请求查询商品信息,服务器以 JSON 格式返回数据,前端 jQuery 代码接收到数据后生成商品编辑界面的分类下拉列表项。第二步:通过 AJAX 请求查询需要修改的商品信息,服务器以 JSON 格式返回数据,前端 jQuery 代码接收到数据后填充商品信息编辑界面。第三步:管理员修改完成后单击"提交"按钮,通过表单以同步消息发送修改后的数据给服务器保存。

在商品查询结果页面单击商品对应的"编辑"超链接,弹出商品信息编辑界面,如图 9-9 所示。修改成功后显示商品的查询结果页面。

图 9-9　商品信息编辑界面

编辑商品信息后单击"提交"按钮，如果信息输入不完整，会给出提示，如图 9-10 所示。

图 9-10　修改商品校验失败界面

9.4.2　任务实现

【任务 9-12】完成方法 GoodsDAO.findById(int id)

编辑 store.dao 包下的 GoodsDAO，添加 findById(int id)方法。GoodsDAO 用来封装对 t_goods

表的操作。findById(int id)方法根据参数 id 查询分类表 t_goods，将结果封装成 Goods 对象返回，代码如下所示。

```
public Goods findById(int id) throws Exception {
    Goods g=null;
    Connection con= null;
    PreparedStatement ps=null;
    ResultSet rs=null;
    try {
        Class.forName(DB.JDBC_DRIVER);
        con=DriverManager.getConnection(DB.JDBC_URL,DB.JDBC_USER,DB.JDBC_PASSWORD);
        ps=con.prepareStatement("select * from t_goods where id=? ");
        ps.setInt(1,id);
        rs=ps.executeQuery();
        if (rs.next()){
            g=new Goods();
            g.setId(rs.getInt("id"));
            g.setName(rs.getString("name"));
            g.setSpecs(rs.getString("specs"));
            g.setSn(rs.getString("sn"));
            g.setPrice(rs.getDouble("price"));
            g.setStock(rs.getInt("stock"));
            g.setCategoryId(rs.getInt("category_id"));
        }
    } catch (Exception e) {
        e.printStackTrace();
        throw new Exception("数据库异常:"+e.getMessage());
    }finally {
        if(rs!=null) rs.close();
        if(ps!=null) ps.close();
        if(con!=null) con.close();
    }
    return g;
}
```

【任务 9-13】完成方法 GoodsDAO.update(Goods g)

编辑 store.dao 包下的 GoodsDAO 类，添加 update(Goods g)方法。

GoodsDAO 用来封装对 t_goods 表的操作。update(Goods g)方法根据参数 g 的 id 将其属性值更新到数据库对应的记录中，代码如下所示。

```
public void update(Goods g) throws Exception {
    Connection con= null;
    PreparedStatement ps=null;
    try {
        Class.forName(DB.JDBC_DRIVER);
        con=DriverManager.getConnection(DB.JDBC_URL,DB.JDBC_USER,DB.JDBC_PASSWORD);
        ps=con.prepareStatement("update t_goods set name=?,specs=?,sn=?,price=?,stock=?,category_id=? where id=?");
        ps.setString(1,g.getName());
        ps.setString(2,g.getSpecs());
        ps.setString(3,g.getSn());
        ps.setDouble(4,g.getPrice());
```

```
            ps.setInt(5,g.getStock());
            ps.setInt(6,g.getCategoryId());
            ps.setInt(7,g.getId());
            ps.executeUpdate();
        } catch (Exception e) {
            e.printStackTrace();
            throw new Exception("数据库异常:"+e.getMessage());
        }finally {
            if(ps!=null) ps.close();
            if(con!=null) con.close();
        }
    }
```

【任务 9-14】完成 GetGoods.java

在 store.controller 包下新建 Servlet 类 GetGoods，其请求地址为/goods。

GetCategory 的 doGet()方法接收前端的 AJAX 请求，通过请求获取参数商品主键，然后通过 GoodsDAO 查询对应的商品信息，最后以 JSON 格式返回数据，代码如下所示。

```
@WebServlet("/goods")
public class GetGoods extends HttpServlet {
    protected void doGet(HttpServletRequest request, HttpServletResponse response) throws ServletException, IOException {
        int id=Integer.parseInt(request.getParameter("id"));
        try {
            Goods g=new GoodsDAO().findById(id);
            JSONObject jsonObject=new JSONObject();
            jsonObject.put("success",true);
            jsonObject.put("data",g);
            response.setContentType("application/x-json");
            response.getWriter().write(jsonObject.toJSONString());
            response.getWriter().flush();
            response.getWriter().close();
        } catch (Exception e) {
            e.printStackTrace();
            JSONObject jsonObject=new JSONObject();
            jsonObject.put("success",false);
            jsonObject.put("msg",e.getMessage());
            response.setContentType("application/x-json");
            response.getWriter().write(jsonObject.toJSONString());
            response.getWriter().flush();
            response.getWriter().close();
        }
    }
}
```

【任务 9-15】完成 EditGoods.java

在 store.controller 包下新建 Servlet 类 EditGoods，其请求地址为/goods/edit。

EditGoods 接收商品更新表单的 post 请求，从请求参数中获取更新数据并将其封装成 Goods 对象，然后调用 GoodsDAO 的 update(Goods g)方法更新数据，完成后转发到资源路径/goods/list，该资源路径会返回商品查询结果页面。代码如下所示。

```
@WebServlet("/goods/edit")
public class EditGoods extends HttpServlet {
```

```java
        protected void doPost(HttpServletRequest request, HttpServletResponse
response) throws ServletException, IOException {
            Goods g=new Goods();
            g.setId(Integer.parseInt(request.getParameter("id")));
            g.setName(request.getParameter("name"));
            g.setSpecs(request.getParameter("specs"));
            g.setSn(request.getParameter("sn"));
            g.setPrice(Double.parseDouble(request.getParameter("price")));
            g.setCategoryId(Integer.parseInt(request.getParameter("category_id")));

            try {
                new GoodsDAO().update(g);
                request.getRequestDispatcher("/goods/list").forward(request,response);
            } catch (Exception e) {
                e.printStackTrace();
                throw new RuntimeException(e.getMessage());
            }
        }
    }
```

【任务 9-16】完成视图部分

商品信息编辑界面通过"编辑"超链接触发编辑模态框。表单提交前需要进行数据验证，验证通过后通过 jQuery 代码提交，因此"提交"按钮的类型是 button，而不是 submit，代码如下所示。

```html
<div class="modal fade" id="editGoods" tabindex="-1">
    <div class="modal-dialog">
        <div class="modal-content">
            <div class="modal-header">
                <h5 class="modal-title">编辑商品</h5>
                <button type="button" class="close" data-dismiss="modal">
                    <span>&times;</span>
                </button>
            </div>
            <div class="modal-body">
                <form id="editGoodsForm" action="/goods/edit" method="post">
                    <div class="form-group row">
                        <label class="col-sm-2 col-form-label">商品</label>
                        <div class="col-sm-10">
                            <input type="hidden" name="id">
                            <input type="text" name="name" class="form-control" placeholder="商品名">
                        </div>
                    </div>
                    <div class="form-group row">
                        <label class="col-sm-2 col-form-label">规格</label>
                        <div class="col-sm-10">
                            <input type="text" name="specs" class="form-control"placeholder="规格">
                        </div>
                    </div>
```

```html
                        <div class="form-group row">
                            <label class="col-sm-2 col-form-label">条形码</label>
                            <div class="col-sm-10">
                                <input type="text" name="sn" class="form-control" placeholder="条形码">
                            </div>
                        </div>
                        <div class="form-group row">
                            <label class="col-sm-2 col-form-label">售价</label>
                            <div class="col-sm-10">
                                <div class="input-group mb-3">
                                    <input type="number" name="price" class="form-control" placeholder="售价">
                                    <div class="input-group-append">
                                        <span class="input-group-text">元</span>
                                    </div>
                                </div>
                            </div>
                        </div>
                        <div class="form-group row">
                            <label class="col-sm-2 col-form-label">分类</label>
                            <div class="col-sm-10">
                                <select name="category_id"class="form-control">
                                    <option>请选择</option>
                                </select>
                            </div>
                        </div>
                        <div class="text-danger d-none alert">
                        </div>
                    </form>
                </div>
                <div class="modal-footer">
                    <button type="button" class="btn btn-secondary" data-dismiss="modal">关闭</button>
                    <button id="editButton" type="button" class="btn btn-primary">提交</button>
                </div>
            </div>
        </div>
    </div>
```

在实现商品查询结果页面时循环输出表格的行数据,每一行都有对应的"编辑"超链接,每个超链接都含有一个自定义属性 goods-id,该属性的值为对应的商品主键。"编辑"超链接代码如下。

```html
<a class="btn btn-sm btn-warning text-white" href="#"
    data-toggle="modal" data-target="#editGoods" goods-id="${g.id}">
    <i class="bi-gear"></i>编辑</a>
```

通过 jQuery 代码 $("a[goods-id]").click()可以选中页面中所有的"编辑"超链接,并且为它们绑定单击事件。在单击事件处理代码中通过$.getJson()向资源/goods 发送 AJAX 请求,在接收到返回值后使用返回的 JSON 数据为模态框中的控件赋值，代码如下。

```javascript
$("a[data-target='#editGoods']").click(function () {
    let goods_id = $(this).attr("goods-id");
    $.getJSON("/categories", function (result) {
        let list = result.data;
        $("#editGoodsForm select[name=category_id]").empty();
        $("#editGoodsForm select[name=category_id]").append('<option value=''>请选择</option>');
        list.forEach(({id, name}) => {
            $("#editGoodsForm select[name=category_id]").append('<option value='\${id}'>\${name}</option>');//避免$和EL表达式冲突，使用转义字符
        });

        $.getJSON("/goods", {id: goods_id}, function (result) {
            $("#editGoodsForm").find("input[name=id]").val(result.data.id);
            $("#editGoodsForm").find("input[name=name]").val(result.data.name);
            $("#editGoodsForm").find("input[name=specs]").val(result.data.specs);
            $("#editGoodsForm").find("input[name=sn]").val(result.data.sn);
            $("#editGoodsForm").find("input[name=price]").val(result.data.price);
            $("#editGoodsForm").find("select[name=category_id]").val(result.data.categoryId);
        });
    });
});
```

这里需要注意的是，异步查询商品信息的操作放在了异步查询分类的回调函数中，读者可以思考一下为什么。

修改完成后单击"提交"按钮，需要对输入的数据进行数据校验。通过 jQuery 代码 $("#editButton").click()为"提交"按钮绑定单击事件，在事件处理代码中依次对必填数据进行校验，如果验证失败，通过 jQuery 找到表单中的警告信息显示层，并在其中显示信息。验证通过则提交表单。代码如下所示。

```javascript
$("#editButton").click(function () {
    let error = "";
    let editForm = $("#editGoodsForm");

    if (editForm.find("input[name=name]").val() == "") {
        error += "请输入商品名<br>";
    }
    if (editForm.find("input[name=specs]").val() == "") {
        error += "请输入商品规格<br>";
    }
    if (editForm.find("input[name=sn]").val() == "") {
        error += "请输入商品条形码<br>";
    }
    if (editForm.find("input[name=price]").val() == "") {
        error += "请输入商品价格<br>";
    }
```

```
        }
        if (editForm.find("select[name=category_id]").val() == "") {
            error += "请选择分类<br>";
        }
        if (error != "") {
            editForm.find("div[class~=alert]").html(error);
            editForm.find("div[class~=alert]").removeClass("d-none");
        } else {
            editForm.find("div[class~=alert]").addClass("d-none");
            editForm.submit();
        }
    });
```

本章习题

1. 结合项目代码,实现查询商品功能。
2. 结合项目代码,实现添加商品功能。
3. 结合项目代码,实现删除商品功能。
4. 结合项目代码,实现修改商品功能。

第10章
进货管理模块实现

本章目标

- 理解进货管理需求
- 理解进货数据表设计
- 理解进货管理各功能时序图
- 掌握 Java 开发技巧
- 掌握 Bootstrap 开发技巧
- 掌握 jQuery 开发技巧

通过前一章的学习，读者已经实现了商品管理模块的功能。本章通过实现超市管理系统的进货管理模块功能，帮助读者加深对 Bootstrap、jQuery 及 Servlet 等知识点的理解，锻炼使用编程语言解决实际问题的能力，强化基于 MVC 模式的编程思维。

10.1 添加进货记录

仔细阅读"添加进货记录"用例描述要求，理解时序图的含义，按要求完成编码工作。

10.1.1 任务需求

添加进货记录用例描述如表 10-1 所示。

表 10-1　添加进货记录用例描述

用例名称	添加进货记录
编号	13
参与者	管理员
简要说明	管理员添加一条进货记录
基本事件流	1. 管理员在商品查询结果页面单击需要进货的商品对应的"进货"超链接 2. 系统显示商品进货界面 3. 管理员在进货界面编辑进货记录后单击"提交"按钮 4. 系统保存进货记录数据 5. 系统返回商品查询页面
其他事件流	无
异常事件流	返回异常页面，提示错误信息
前置条件	管理员已经登录系统
后置条件	进货记录保存到数据库

添加进货记录功能时序图如图 10-1 所示。

时序图中各参与者说明如下。

goods.jsp：视图层。单击"进货"超链接会发送一个 AJAX 请求，请求后台查询所有的供应商信息，显示添加进货记录表单，并根据返回的 JSON 数据生成供应商列表的内容。

GetSuppliers：控制层，位于 store.controller 包中，是一个 Servlet。其请求地址为/suppliers，处理来自 goods.jsp 的异步 get 请求，调用 SupplierDAO 的 findSuppliers(String name)方法查询供应商信息，操作完成后以 JSON 格式返回供应商信息。

AddRestock：控制层，位于 store.controller 包中，是一个 Servlet。其请求地址为/restock/add，处理来自 goods.jsp 的 post 请求，用于获取请求参数中的进货记录信息并将其封装成 Restock 对象，调用 RestockDAO 的 save(Restock r)方法添加进货记录，操作完成后返回商品查询页面。

SupplierDAO：数据访问模型层。其 findSuppliers(String name)方法根据供应商名模糊查找 t_supplier 表，将符合条件的供应商信息返回。

第 10 章
进货管理模块实现

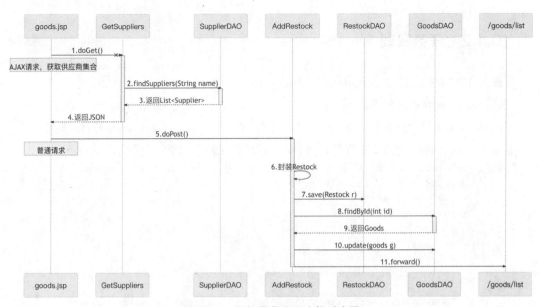

图 10-1　添加进货记录功能时序图

RestockDAO：数据访问模型层。其 save(Restock r)方法将一个 Restock 对象的值通过 JDBC 添加到 t_restock 表中。

GoodsDAO：数据访问模型层。其 update(Goods g)方法将一个 Goods 对象的值通过 JDBC 更新到 t_goods 表中。

/goods/list：商品查询控制层对应的地址，请求该地址最终会返回商品查询视图层。

在商品查询结果页面单击商品对应的"进货"超链接，弹出进货信息添加界面，如果进货信息不完整，会提示错误信息，如图 10-2 所示。

图 10-2　添加进货记录界面

在查询供应商输入框中输入供应商名，单击后面的"查询"按钮，可以异步模糊查询供应商信息，提高交互效率。

207

10.1.2 任务实现

【任务 10-1】完成方法 SupplierDAO.findSuppliers(String name)

编辑 store.controller 包下的 SupplierDAO 类，添加 findSuppliers(String name)方法。

SupplierDAO 用来封装对 t_supplier 表的操作，findSuppliers(String name)方法根据参数值对 name 字段进行模糊查询，并将符合条件的结果封装成集合返回，代码如下所示。

```java
public List<Supplier> findSuppliers(String name) throws Exception {
    List<Supplier> list=new ArrayList<>();
    Connection con= null;
    PreparedStatement ps=null;
    ResultSet rs=null;
    try {
        Class.forName(DB.JDBC_DRIVER);
        con=DriverManager.getConnection(DB.JDBC_URL,DB.JDBC_USER,DB.JDBC_PASSWORD);
        ps=con.prepareStatement("select * from t_supplier where name like ? ");
        ps.setString(1,"%"+name+"%");
        rs=ps.executeQuery();
        while (rs.next()){
            Supplier s=new Supplier();
            s.setId(rs.getInt("id"));
            s.setName(rs.getString("name"));
            s.setTel(rs.getString("tel"));
            s.setContacts(rs.getString("contacts"));
            s.setInfo(rs.getString("info"));
            list.add(s);
        }
    } catch (Exception e) {
        e.printStackTrace();
        throw new Exception("数据库异常:"+e.getMessage());
    }finally {
        if(rs!=null) rs.close();
        if(ps!=null) ps.close();
        if(con!=null) con.close();
    }
    return list;
}
```

【任务 10-2】完成 GetSuppliers.java

在 store.controller 包下新建 Servlet 类 GetSuppliers，其请求地址为/suppliers，注意不要与 GetSupplier 类混淆，GetSupplier 类是根据供应商主键查询供应商，GetSuppliers 类是根据供应商名查询供应商集合。

GetCategories 的 doGet()方法接收前端的 AJAX 请求，然后通过 SupplierDAO 查询供应商信息集合，最后以 JSON 格式返回数据。代码如下所示。

```java
@WebServlet("/suppliers")
public class GetSuppliers extends HttpServlet {
    protected void doGet(HttpServletRequest request, HttpServletResponse response) throws ServletException, IOException {
```

```
            String name=request.getParameter("name");
            if(name==null) name="";
            try {
                List<Supplier> list=new SupplierDAO().findSuppliers(name);
                JSONObject jsonObject=new JSONObject();
                jsonObject.put("success",true);
                jsonObject.put("data",list);
                response.setContentType("application/x-json");
                response.getWriter().write(jsonObject.toJSONString());
                response.getWriter().flush();
                response.getWriter().close();
            } catch (Exception e) {
                e.printStackTrace();
                JSONObject jsonObject=new JSONObject();
                jsonObject.put("success",false);
                jsonObject.put("msg",e.getMessage());
                response.setContentType("application/x-json");
                response.getWriter().write(jsonObject.toJSONString());
                response.getWriter().flush();
                response.getWriter().close();
            }
        }
    }
```

【任务 10-3】完成方法 RestockDAO.save(Restock r)

在 store.dao 包下新建 RestockDAO 类，并在其中添加方法 save(Restock r)。

RestockDAO 用来封装对 t_restock 表的操作，save(Restock r)方法将 Restock 类型参数 r 的属性值保存成 t_restock 表中的一条记录，代码如下所示。

```
public void save(Restock r) throws Exception {
    Connection con = null;
    PreparedStatement ps = null;
    try {
        Class.forName(DB.JDBC_DRIVER);
        con = DriverManager.getConnection(DB.JDBC_URL, DB.JDBC_USER, DB.JDBC_PASSWORD);
        ps = con.prepareStatement("insert into t_restock value (null ,?,?,?,?,?)");
        ps.setInt(1, r.getGoodsId());
        ps.setInt(2, r.getSupplierId());
        ps.setDouble(3, r.getPrice());
        ps.setInt(4, r.getStock());
        ps.setString(5, r.getTradeDate());
        ps.executeUpdate();
    } catch (Exception e) {
        e.printStackTrace();
        throw new Exception("数据库异常:" + e.getMessage());
    } finally {
        if (ps != null) ps.close();
        if (con != null) con.close();
    }
}
```

【任务 10-4】完成 AddRestock.java

在 store.controller 包下新建 Servlet 类 AddRestock，其请求地址为/restock/add。

AddRestock 用来接收 addRestockForm 表单的请求数据。它将数据封装成一个 Restock 对象，然后调用 RestockDAO 的 save()方法进行保存。需要注意的是，因为是添加操作，被添加的数据还没有主键，所以在封装被添加对象时不需要为 id 赋值。保存了 Restock 对象后，还需要将商品信息查询出来，并且在修改商品库存数量后更新商品信息。操作成功后转发到资源地址/goods/list，代码如下所示。

```java
@WebServlet("/restock/add")
public class AddRestock extends HttpServlet {
    protected void doPost(HttpServletRequest request, HttpServletResponse response) throws ServletException, IOException {
        Restock r = new Restock();
        r.setGoodsId(Integer.parseInt(request.getParameter("goods_id")));
        r.setPrice(Double.parseDouble(request.getParameter("price")));
        r.setStock(Integer.parseInt(request.getParameter("stock")));
        r.setSupplierId(Integer.parseInt(request.getParameter("supplier_id")));
        r.setTradeDate(new SimpleDateFormat("yyyy-MM-dd").format(new Date()));
        try {
            new RestockDAO().save(r);
            GoodsDAO goodsDAO = new GoodsDAO();
            Goods g = goodsDAO.findById(r.getGoodsId());
            g.setStock(r.getStock() + g.getStock());
            goodsDAO.update(g);
            request.getRequestDispatcher("/goods/list?name="+g.getName()).forward(request,response);
            response.getWriter().close();
        } catch (Exception e) {
            e.printStackTrace();
            throw new RuntimeException(e.getMessage());
        }
    }
}
```

【任务 10-5】完成视图部分

添加进货记录信息输入界面通过 goods.jsp 页面中的"进货"超链接触发模态框生成，代码如下所示。

```html
<a class="btn btn-sm btn-primary text-white" href="#"
    data-toggle="modal" data-target="#addRestock" goods-id="${g.id}" goods-name="${g.name}">
    <i class="bi-gear"></i>进货</a>
```

代码中 data-toggle="modal"用来指定 Bootstrap 的触发行为是模态框，data-target="#addRestock" 用来指定要显示的模态框组件 id，对应下面代码中的 id="addRestock"。

```html
<div class="modal fade" id="addRestock" tabindex="-1">
    <div class="modal-dialog">
        <div class="modal-content">
            <div class="modal-header">
                <h5 class="modal-title">进货</h5>
                <button type="button" class="close" data-dismiss="modal">
```

```html
                                    <span>&times;</span>
                                </button>
                            </div>
                            <div class="modal-body">
                                <form id="addRestockForm" action="/restock/add" method="post">
                                    <div class="form-group row">
                                        <label class="col-sm-2 col-form-label">商品</label>
                                        <div class="col-sm-10">
                                            <input type="hidden" name="goods_id">
                                            <input type="text" name="name" class="form-control" value="" disabled>
                                        </div>
                                    </div>
                                    <div class="form-group row">
                                        <label class="col-sm-2 col-form-label">单价</label>
                                        <div class="col-sm-10">
                                            <div class="input-group mb-3">
                                                <input type="number" name="price" class="form-control" placeholder="进货价">
                                                <div class="input-group-append">
                                                    <span class="input-group-text">元</span>
                                                </div>
                                            </div>
                                        </div>
                                    </div>

                                    <div class="form-group row">
                                        <label class="col-sm-2 col-form-label">数量</label>
                                        <div class="col-sm-10">
                                            <input type="number" name="stock" class="form-control" placeholder="进货数量">
                                        </div>
                                    </div>
                                    <div class="form-group row">
                                        <label class="col-sm-2 col-form-label">供应商</label>
                                        <div class="col-sm-5">
                                            <select name="supplier_id"class="form-control">
                                                <option>请选择</option>
                                            </select></div>
                                        <div class="col-5">
                                            <div class="input-group mb-3">
                                                <input id="supplier_name" type="text" class="form-control" placeholder="查询供应商">
                                                <div class="input-group-append">
                                                    <button id="search_supplier" class="btn btn-success" type="button"><i class="bi-search"></i></button>
                                                </div>
                                            </div>
                                        </div>
                                    </div>
```

```
                        <div class="text-danger d-none alert">
                        </div>

                    </form>
                </div>
                <div class="modal-footer">
                    <button type="button" class="btn btn-secondary" data-dismiss="modal">关闭</button>
                    <button id="addRestockButton" type="button" class="btn btn-primary">提交</button>
                </div>
            </div>
        </div>
    </div>
```

添加进货记录表单中的"供应商"下拉列表默认只有一个"请选择"选项,当管理员单击"进货"超链接时,需要通过 AJAX 请求资源地址/suppliers 查询供应商集合,并根据返回结果生成下拉列表项,代码如下所示。

```
$("a[data-target='#addRestock']").click(function () {
    let goods_id = $(this).attr("goods-id");
    let goods_name = $(this).attr("goods-name");
    $("#addRestockForm input[name=goods_id]").val(goods_id);
    $("#addRestockForm input[name=name]").val(goods_name);
    let select_supplier = $("#addRestock select[name=supplier_id]");
    $.getJSON("/suppliers", function (result) {
        select_supplier.empty();
        select_supplier.append('<option value="">请选择</option>');
        result.data.forEach(({id, name}) => {
            select_supplier.append('<option value="\${id}">\${name}</option>');
        });
    });
});
```

代码 `<button id="search_supplier" class="btn btn-success" type="button">` 中为查询供应商按钮设置了 id 值,通过 jQuery 代码为该按钮绑定了 click 事件处理代码,当按钮被单击时异步查询供应商信息,并重新生成供应商下拉列表项,代码如下所示。

```
$("#search_supplier").click(function () {
    let select_supplier = $("#addRestock select[name=supplier_id]");
    $.getJSON("/suppliers", {name: $("#supplier_name").val()}, function (result) {
        select_supplier.empty();
        select_supplier.append('<option value="">请选择</option>');
        result.data.forEach(({id, name}) => {
            select_supplier.append('<option value="\${id}">\${name}</option>');
        });
    });
});
```

代码 `<button id="addRestockButton" type="button" class="btn btn-primary">` 中为"提交"按钮设置了 id 值,通过 jQuery 代码为该按钮绑定了 click 事件处理代码,当"提交"按钮被单击时,通过 CSS 选择器找到 id="addRestockForm" 表单里面需要进行非空验证的控件进行数据验证，代码如下所示。

```javascript
$("#addRestockButton").click(function () {
    let error = "";
    let addRestockForm = $("#addRestockForm");
    if (addRestockForm.find("input[name=price]").val() == "") {
        error += "请输入商品价格<br>";
    }
    if (addRestockForm.find("input[name=stock]").val() == "") {
        error += "请输入数量<br>";
    }
    if (addRestockForm.find("select[name=supplier_id]").val() == "") {
        error += "请选择供应商<br>";
    }

    if (error != "") {
        addRestockForm.find("div[class~=alert]").html(error);
        addRestockForm.find("div[class~=alert]").removeClass("d-none");
    } else {
        addRestockForm.find("div[class~=alert]").addClass("d-none");
        addRestockForm.submit();
    }

});
```

10.2 查询进货记录

仔细阅读"查询进货记录"用例描述要求，理解时序图的含义，按要求完成编码工作。

10.2.1 任务需求

查询进货记录用例描述如表 10-2 所示。

表 10-2　查询进货记录用例描述

用例名称	查询进货记录
编号	14
参与者	管理员
简要说明	管理员输入条件查询进货记录
基本事件流	1. 管理员在查询表单中输入查询条件，单击"查询"按钮 2. 系统显示符合条件的进货记录
其他事件流	无
异常事件流	返回异常页面，提示错误信息
前置条件	管理员已经登录系统
后置条件	无

查询进货记录功能时序图如图 10-3 所示。

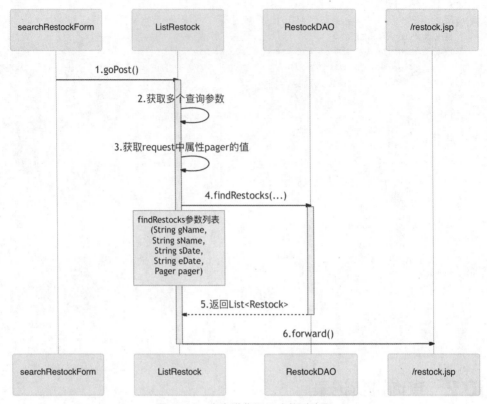

图 10-3　查询进货记录功能时序图

时序图中各参与者说明如下。

searchRestockForm：视图层，放置于 restock.jsp 里面的表单，提交方式为 post，提交地址为/restock/list，包含控件商品名 goodsName、供应商名 supplierName、开始时间 startDate 和结束时间 endDate。

ListRestock：控制层，位于 store.controller 包中，是一个 Servlet。请求地址为/restock/list，处理来自 searchRestockForm 的 post 请求，用于获取查询条件商品名 goodsName、供应商名 supplierName、开始时间 startDate 和结束时间 endDate，并调用 RestockDAO 的 findRestocks(String gName, String sName, String sDate, String eDate, Pager pager)方法进行分页查询，操作完成后返回进货记录查询视图。当单击"上一页"或"下一页"按钮进行数据查询时，超链接发送的是 get 方式的请求，由于其处理逻辑和表单 post 方式请求完全相同，因此 ListRestocks 中的 doGet()方法直接调用 doPost()方法即可。

RestockDAO：数据访问模型层。其 findRestocks(String gName, String sName, String sDate, String eDate, Pager pager)方法进行分页查询。

/restock.jsp：进货记录查询页面的地址。请求该地址最终会返回进货记录查询视图层。

在进货记录查询主页面的条件输入框中输入查询条件并单击"查询"按钮，分页显示满足条件的查询结果，如图 10-4 所示。

图 10-4　查询进货记录主页面

10.2.2　任务实现

【任务10-6】完成方法 RestockDAO.findRestocks(String gName, String sName, String sDate, String eDate, Pager pager)

编辑 store.controller 包下的 RestockDAO 类，添加 findRestocks(String gName, String sName, String sDate, String eDate, Pager pager)方法。

RestockDAO 用来封装对 t_restock 表的操作，其 findRestocks()方法进行分页查询，其中的参数 gName 为商品名、sName 为供应商名、sDate 为开始时间、eDate 为结束时间、pager 为分页帮助对象。显示进货记录时需要显示商品名和供应商名，t_restock 表中没有相关信息，因此要联合查询 t_goods 表和 t_supplier 表，代码如下所示。

```
public List<Restock> findRestockes(String gName, String sName, String sDate, String eDate, Pager pager) throws Exception {
    List<Restock> list = new ArrayList<>();
    Connection con = null;
    PreparedStatement ps = null;
    ResultSet rs = null;
    try {
        Class.forName(DB.JDBC_DRIVER);
        con = DriverManager.getConnection(DB.JDBC_URL, DB.JDBC_USER, DB.JDBC_PASSWORD);
        ps = con.prepareStatement("select count(r.id) as total from t_restock as r " +
                "inner join t_supplier as s on r.supplier_id=s.id " +
                "inner join t_goods as g on r.goods_id=g.id " +
                "where g.name like ? and s.name like ? and trade_date between ? and ?");
        ps.setString(1, "%" + gName + "%");
        ps.setString(2, "%" + sName + "%");
        ps.setString(3, sDate);
        ps.setString(4, eDate);
        rs=ps.executeQuery();
        if(rs.next()){
```

```java
                    pager.setTotal(rs.getInt("total"));
                }
                
                ps = con.prepareStatement("select r.*,s.name as sname,g.name as gname from t_restock as r " +
                        "inner join t_supplier as s on r.supplier_id=s.id " +
                        "inner join t_goods as g on r.goods_id=g.id " +
                        "where g.name like ? and s.name like ? and trade_date between ? and ? limit ?,?");
                ps.setString(1, "%" + gName + "%");
                ps.setString(2, "%" + sName + "%");
                ps.setString(3, sDate);
                ps.setString(4, eDate);
                ps.setInt(5,(pager.getCurrentPage()-1)*pager.getPageSize());
                ps.setInt(6,pager.getPageSize());
                rs = ps.executeQuery();
                while (rs.next()) {
                    Restock r = new Restock();
                    r.setId(rs.getInt("id"));
                    r.setGoodsId(rs.getInt("goods_id"));
                    r.setSupplierId(rs.getInt("supplier_id"));
                    r.setPrice(rs.getDouble("price"));
                    r.setStock(rs.getInt("stock"));
                    r.setTradeDate(rs.getString("trade_date"));
                    r.setGoodsName(rs.getString("gname"));
                    r.setSupplier(rs.getString("sname"));
                    list.add(r);
                }
        } catch (Exception e) {
            e.printStackTrace();
            throw new Exception("数据库异常:" + e.getMessage());
        } finally {
            if (rs != null) rs.close();
            if (ps != null) ps.close();
            if (con != null) con.close();
        }
        return list;
    }
```

【任务 10-7】完成 ListRestock.java

在 store.controller 包下新建 Servlet 类 ListRestock,其请求地址为/restock/list。

ListRestock 的 doPost()方法用来处理进货记录查询表单的请求,doGet()方法用来处理分页按钮的请求,由于两者的逻辑相同,因此在 doGet()方法中直接调用 doPost()方法即可。ListRestock 从请求参数中得到查询条件 goodsName、supplierName、startDate、endDate,从请求对象的属性中得到分页对象 pager,将它们作为参数对 RestockDAO 的 findRestocks()方法进行调用,得到分页后的查询结果 restockList。查询完毕后通过 request 对象的属性将 pager 和 restockList 两个对象转发到 restock.jsp 页面,代码如下所示。

```java
@WebServlet("/restock/list")
public class ListRestock extends HttpServlet {
    protected void doPost(HttpServletRequest request, HttpServletResponse response) throws ServletException, IOException {
```

```java
            String goods=request.getParameter("goodsName");
            String supplier=request.getParameter("supplierName");
            String startDate=request.getParameter("startDate");
            String endDate=request.getParameter("endDate");
            if(goods==null) goods="";
            if(supplier==null) supplier="";
            String now=new SimpleDateFormat("yyyy-MM-dd").format(new Date());
            if(startDate==null || "".equals(startDate)) startDate=now;
            if(endDate==null || "".equals(endDate)) endDate=now;
            Pager pager = (Pager) request.getAttribute("pager");
            try {
                List<Restock> restockList=new RestockDAO().findRestockes(goods,supplier,startDate,endDate,pager);
                request.setAttribute("restockList",restockList);
                request.getRequestDispatcher("/restock.jsp").forward(request,response);
            } catch (Exception e) {
                e.printStackTrace();
                throw new RuntimeException(e.getMessage());
            }
        }

        protected void doGet(HttpServletRequest request, HttpServletResponse response) throws ServletException, IOException {
            doPost(request,response);
        }
    }
```

【任务10-8】完成视图部分

查询进货记录表单 searchRestockForm 的实现代码位于 restock.jsp 中，以 post 方式提交到/restock/list，提交参数 goodsName 为商品名、supplierName 为供应商名、startDate 为开始时间、endDate 为结束时间，代码如下所示。

```html
<form class="form-inline" action="/restock/list" method="post">
    <input type="text" name="goodsName" class="form-control mr-1" placeholder="请输入商品">
    <input type="text" name="supplierName" class="form-control mr-1" placeholder="请输入供应商">
    <label class="mr-1">从</label>
    <input type="date" name="startDate" class="form-control mr-1" />
    <label class="mr-1">到</label>
    <input type="date" name="endDate" class="form-control mr-1" />
    <button type="submit" class="btn btn-success"><i class="bi-search"></i>查询</button>
</form>
```

ListRestock 会通过 request 作用域返回两个对象，分别是分页查询结果集 restockList 和分页对象 pager，在 restock.jsp 中通过循环 restockList 生成结果表格，代码如下所示。

```html
<table class="table table-hover table-bordered table-striped text-center">
    <thead class="thead-dark">
    <tr>
        <th >序号</th>
        <th >商品名</th>
```

```
            <th >供应商</th>
            <th >进货日期</th>
            <th >进价</th>
            <th >数量</th>
        </tr>
        </thead>
        <tbody>
        <c:forEach items="${requestScope.restockList}" var="r" varStatus="vs">
            <tr>
                <td>${vs.count}</td>
                <td>${r.goodsName}</td>
                <td>${r.supplier}</td>
                <td>${r.tradeDate}</td>
                <td>${r.price}</td>
                <td>${r.stock}</td>
            </tr>
        </c:forEach>
        </tbody>
    </table>
```

分页按钮部分代码如下所示，其中超链接请求/restock/list 进行分页查询；传递参数 currentPage 表示希望跳转到第几页，从分页对象 pager 中获取；多个查询条件从请求参数中获取。

```
    <nav>
        <ul class="pagination justify-content-between">
            <li class="page-item">
                <a class="page-link rounded-pill" href="/restock/list?currentPage=${requestScope.pager.previous}&goodsName=${param.goodsName}&supplierName=${param.supplierName}&startDate=${param.startDate}&endDate=${param.endDate}" tabindex="-1">上一页</a>
            </li>
            <li class="page-item"><a class="page-link rounded-pill">总共${requestScope.pager.pageCount}页，当前第${requestScope.pager.currentPage}页</a></li>
            <li class="page-item">
                <a class="page-link rounded-pill"href="/restock/list?currentPage=${requestScope.pager.next}&goodsName=${param.goodsName}&supplierName=${param.supplierName}&startDate=${param.startDate}&endDate=${param.endDate}">下一页</a>
            </li>
        </ul>
    </nav>
```

本章习题

1. 结合项目代码，实现添加进货记录功能。
2. 结合项目代码，实现查询进货记录功能。

第11章
销售模块实现

本章目标

- 理解销售模块需求
- 理解销售数据表设计
- 理解销售模块各功能时序图
- 掌握 Java 开发技巧
- 掌握 Bootstrap 开发技巧
- 掌握 jQuery 开发技巧

通过前一章的学习，读者已经实现了进货管理模块的功能。本章通过实现超市管理系统的销售模块功能，帮助读者加深对 Bootstrap、jQuery 及 Servlet 等知识点的理解，锻炼使用编程语言解决实际问题的能力，强化基于 MVC 模式的编程思维。

11.1 添加销售记录

仔细阅读"添加销售记录"用例描述要求，理解时序图的含义，按要求完成编码工作。

11.1.1 任务需求

添加销售记录用例描述如表 11-1 所示。

表 11-1 添加销售记录用例描述

用例名称	添加销售记录
编号	15
参与者	管理员
简要说明	管理员添加一条销售记录
基本事件流	1. 管理员在销售管理页面单击"添加销售记录"按钮 2. 系统显示添加销售记录界面 3. 管理员通过商品编码查找商品信息 4. 管理员修改销售商品数量 5. 管理员编辑完销售明细后单击"提交"按钮 6. 系统保存销售记录和销售明细数据，更新库存 7. 系统返回销售管理页面
其他事件流	无
异常事件流	返回异常页面，提示错误信息
前置条件	管理员已经登录系统
后置条件	1. 销售记录保存到销售记录表 2. 销售明细保存到销售明细表 3. 商品库存更新到商品表

添加销售记录功能时序图如图 11-1 所示。

时序图中各参与者说明如下。

sale.jsp：视图层。单击"添加销售记录"按钮显示添加销售记录表单。表单中含有商品条形码输入框，输入条形码可以异步查询对应的商品信息并根据返回的 JSON 数据生成销售明细。

GetGoodsBySn：控制层，位于 store.controller 包中，是一个 Servlet。其请求地址为/goods/sn，处理来自 sale.jsp 的异步 get 请求，调用 GoodsDAO 的 findBySn(String sn)方法查询供应商信息，操作完成后以 JSON 格式返回分类信息。

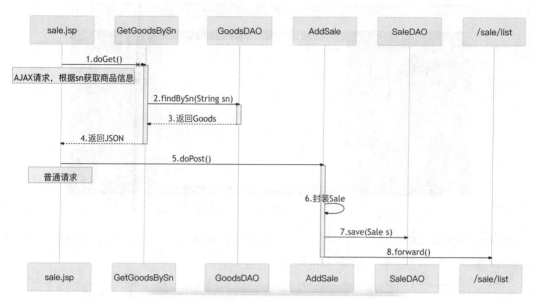

图 11-1 添加销售记录功能时序图

AddSale：控制层，位于 store.controller 包中，是一个 Servlet。其请求地址为/sale/add，处理来自 sale.jsp 的 post 请求，用于获取请求参数中的销售明细信息并将其封装成 Sale 对象，调用 SaleDAO 的 save(Sale s)方法添加销售记录，操作完成后返回销售记录查询页面。

GoodsDAO：数据访问模型层。其 findBySn(String sn)方法根据条形码模糊查找 t_goods 表，将符合条件的商品信息返回。

SaleDAO：数据访问模型层。其 save(Sale s)方法将一个 Sale 对象的值通过 JDBC 添加到 t_sale 表中，添加对应的销售明细到 t_sale_item 销售明细表，同时还需要更新商品库存到 t_goods 表。

/sale/list：销售记录查询控制层对应的地址，请求该地址最终会返回销售记录查询视图层。

在销售管理页面单击"添加销售记录"按钮，弹出销售记录添加界面，如图 11-2 所示。

图 11-2 添加销售记录界面

在"商品编码"输入框中输入商品条形码，单击"添加商品"按钮可以增加一条销售明细，在销售明细中可以修改商品数量，如图 11-3 所示。

图 11-3　编辑销售明细界面

销售明细编辑完成后单击"提交"按钮，系统保存销售记录，返回销售记录查询页面可以看见刚刚添加的销售记录，如图 11-4 所示。

图 11-4　销售记录添加完成界面

11.1.2　任务实现

【任务 11-1】完成方法 GoodsDAO.findBySn(String sn)

编辑 store.controller 包下的 GoodsDAO 类，添加 findBySn(String sn)方法。

GoodsDAO 用来封装对 t_goods 表的操作，findBySn(String sn)方法根据参数值对 sn 字段进行查询，并将结果封装成 Goods 返回，代码如下所示。

```
public Goods findBySn(String sn) throws Exception {
    Goods g=null;
    Connection con= null;
    PreparedStatement ps=null;
    ResultSet rs=null;
    try {
        Class.forName(DB.JDBC_DRIVER);
        con=DriverManager.getConnection(DB.JDBC_URL,DB.JDBC_USER,DB.JDBC_PASSWORD);
        ps=con.prepareStatement("select * from t_goods where sn=? ");
        ps.setString(1,sn);
        rs=ps.executeQuery();
        if (rs.next()){
```

```
                g=new Goods();
                g.setId(rs.getInt("id"));
                g.setName(rs.getString("name"));
                g.setSpecs(rs.getString("specs"));
                g.setSn(rs.getString("sn"));
                g.setPrice(rs.getDouble("price"));
                g.setStock(rs.getInt("stock"));
                g.setCategoryId(rs.getInt("category_id"));
            }
        } catch (Exception e) {
            e.printStackTrace();
            throw new Exception("数据库异常:"+e.getMessage());
        }finally {
            if(rs!=null) rs.close();
            if(ps!=null) ps.close();
            if(con!=null) con.close();
        }
        return g;
    }
```

【任务 11-2】完成 GetGoodsBySn.java

在 store.controller 包下新建 Servlet 类 GetGoodsBySn,其请求地址为/goods/sn。

GetGoodsBySn 的 doGet()方法接收前端的 AJAX 请求,然后通过 GoodsDAO 查询商品信息,最后以 JSON 格式返回数据。

```
@WebServlet("/goods/sn")
public class GetGoodsBySn extends HttpServlet {
    protected void doGet(HttpServletRequest request, HttpServletResponse response) throws ServletException, IOException {
            String sn=request.getParameter("sn");
            try {
                Goods g=new GoodsDAO().findBySn(sn);
                JSONObject jsonObject=new JSONObject();
                jsonObject.put("success",true);
                jsonObject.put("data",g);
                response.setContentType("application/x-json");
                response.getWriter().write(jsonObject.toJSONString());
                response.getWriter().flush();
                response.getWriter().close();
            } catch (Exception e) {
                e.printStackTrace();
                JSONObject jsonObject=new JSONObject();
                jsonObject.put("success",false);
                jsonObject.put("msg",e.getMessage());
                response.setContentType("application/x-json");
                response.getWriter().write(jsonObject.toJSONString());
                response.getWriter().flush();
                response.getWriter().close();
            }
    }
}
```

【任务 11-3】完成方法 SaleDAO.save(Sale s)

在 store.dao 包下新建 SaleDAO 类,并在其中添加方法 save(Sale s)。

SaleDAO 的 save(sale s)方法需要完成销售记录保存、销售明细保存和商品库存修改，代码如下所示。

```java
public void save(Sale s) throws Exception {
    Connection con = null;
    PreparedStatement ps = null;
    ResultSet rs=null;
    try {
        Class.forName(DB.JDBC_DRIVER);
        con = DriverManager.getConnection(DB.JDBC_URL, DB.JDBC_USER, DB.JDBC_PASSWORD);
        con.setAutoCommit(false);
        ps = con.prepareStatement("insert into t_sale value (null ,?)", Statement.RETURN_GENERATED_KEYS);
        ps.setString(1, s.getTradeTime());
        ps.executeUpdate();
        rs=ps.getGeneratedKeys();
        if(rs.next()){//销售记录生成后添加销售记录项
            StringBuilder sb=new StringBuilder();
            int sale_id=rs.getInt(1);
            s.getItems().forEach(saleItem ->{
                sb.append("insert into t_sale_item value("+sale_id+","+saleItem.getGoodsId()+","+saleItem.getCount()+"); ");
            });

            //修改库存
            s.getItems().forEach(saleItem -> {
                sb.append("update t_goods set stock=(stock- "+saleItem.getCount()+") where id="+saleItem.getGoodsId()+" ; ");
            });
            ps=con.prepareStatement(sb.toString());
            ps.executeUpdate();
        }
        con.commit();
    } catch (Exception e) {
        con.rollback();
        e.printStackTrace();
        throw new Exception("数据库异常:" + e.getMessage());
    } finally {
        if(rs!=null) rs.close();
        if (ps != null) ps.close();
        if (con != null) con.close();
    }
}
```

销售记录添加、销售明细添加，以及商品库存修改等操作要么都做，要么都不做，它们是一个不可分割的操作序列，因此需要通过 JDBC 事务进行处理。

在事务开始时通过 setAutoCommit(false)关闭 JDBC 的自动事务提交，在所有操作完成后通过 commit()方法提交事务，如果操作出现异常，则需要回滚事务。

添加销售明细时需要为其对应的销售记录主键赋值，需要在添加销售记录时获取数据库自动生成的主键，因此在执行添加销售记录的 prepareStatement()方法时需要指定第二参数值

Statement.RETURN_GENERATED_KEYS。通过上述操作后就可以利用 getGeneratedKeys()方法得到销售记录的主键。

【任务 11-4】完成 AddSale.java

在 store.controller 包下新建 Servlet 类 AddSale，其请求地址为/sale/add。

AddSale 用来接收 addSaleForm 表单的请求数据。它将数据封装成一个 Sale 对象，然后调用 SaleDAO 的 save()方法进行保存。销售明细中商品主键和其对应的数量通过数组的方式传递。操作成功后转发到资源地址/sale/list，代码如下所示。

```
@WebServlet("/sale/add")
public class AddSale extends HttpServlet {
    protected void doPost(HttpServletRequest request, HttpServletResponse response) throws ServletException, IOException {
        try {
            String[] goodsIds=request.getParameterValues("goodsId");
            String[] counts=request.getParameterValues("count");
            String now=new SimpleDateFormat("yyyy-MM-dd hh:mm:ss").format(new Date());
            Sale s=new Sale();
            s.setTradeTime(now);
            for (int i=0;i<goodsIds.length;i++){
                SaleItem saleItem=new SaleItem();
                saleItem.setGoodsId(Integer.parseInt(goodsIds[i]));
                saleItem.setCount(Integer.parseInt(counts[i]));
                s.getItems().add(saleItem);
            }
            new SaleDAO().save(s);
            request.getRequestDispatcher("/sale/list").forward(request,response);
        } catch (Exception e) {
            e.printStackTrace();
        }
    }
}
```

【任务 11-5】完成视图部分

添加销售记录信息输入界面通过 sale.jsp 页面中的"添加销售记录"按钮触发模态框生成，代码如下所示。

```
<button type="button" class="btn btn-primary" data-toggle="modal" data-target="#addSale"><i
        class="bi-plus"></i>添加销售记录
</button>
```

代码中 data-toggle="modal"用来指定 Bootstrap 的触发行为是模态框，data-target="#addSale" 用来指定要显示的模态框组件 id，对应下面代码中的 id="addSale"。

```
<div class="modal fade" id="addSale">
    <div class="modal-dialog">
        <div class="modal-content" style="width:max-content">
            <div class="modal-header">
                <h5 class="modal-title">添加销售记录</h5>
                <button type="button" class="close" data-dismiss="modal">
```

```html
                            <span>&times;</span>
                        </button>
                    </div>
                    <form id="addSaleForm"  action="/sale/add" method="post">
                        <div class="modal-body">
                            <div class="form-group row">
                                <label class="col-sm-3 col-form-label">商品编码</label>
                                <div class="col-sm-9">
                                    <div class="input-group">
                                        <input type="text" id="sn" class="form-control"placeholder="请输入条形码">
                                        <div class="input-group-append">
                                            <button id="addSaleItem" class="btn btn-outline-secondary" type="button" >添加商品</button>
                                        </div>
                                    </div>
                                </div>
                            </div>
                            <table class="table table-hover table-striped">
                                <thead>
                                <tr>
                                    <td>条形码</td>
                                    <td>商品名</td>
                                    <td>单价</td>
                                    <td>库存</td>
                                    <td>数量</td>
                                </tr>
                                </thead>
                                <tbody id="saleDetail">

                                </tbody>
                                <tfoot>
                                <tr>
                                    <td colspan="4">
                                        总共<span id="totalCount">0</span>件商品，总价<span id="totalPrice">0</span>元。
                                    </td>
                                </tr>
                                </tfoot>
                            </table>
                            <div class="modal-footer">
                                <button type="button" class="btn btn-default" data-dismiss="modal">关闭</button>
                                <button id="addButton" type="button" class="btn btn-primary">提交</button>
                            </div>
                    </form>
                </div>
            </div>
        </div>
```

当管理员在"商品编码"输入框中输入商品条形码并单击"添加商品"按钮时,AJAX 请求资源地址/goods/sn 查询商品信息,并根据返回结果生成销售明细,代码如下。

```
$("#addSaleItem").click(function () {
    let sn = $("#sn").val();
    $.getJSON("/goods/sn", {sn: sn}, function (result) {
        let goods = result.data;
        if (goods == undefined) {
            window.alert("商品不存在");
        } else {
            $("#saleDetail").append(' <tr>
                <td>\${goods.sn}</td>
                <td>\${goods.name}</td>
                <td><input name="price" disabled="disabled" value="\${goods.price}"></input></td>
                <td>\${goods.stock}</td>
                <td>
                    <input type="hidden" name="goodsId" value="\${goods.id}">
                    <input type="number" name="count" step="1" min="1" max="\${goods.stock}" value="1" contenteditable="false" onchange="sum()">
                </td>
            </tr>');
            sum();
        }
    });
});
```

每一条销售明细中都可以修改商品数量,当明细发生变化时会计算销售总金额,需要使用 sum()方法,代码如下所示。

```
function sum() {
    let totalCount = 0;
    let totalPrice = 0;
    let items = $("#saleDetail").find("tr");
    $.each(items, function (i, item) {
        let price = parseFloat($(item).find("[name=price]").val());
        let count = parseInt($(item).find("[name=count]").val());
        totalCount += count;
        totalPrice += (count * price);
    });

    $("#totalCount").text(totalCount);
    $("#totalPrice").text(Math.ceil(totalPrice*100)/100);
}
```

代码 `<button id="addButton" type="button" class="btn btn-primary">` 中为"提交"按钮设置了 id 值,通过 jQuery 代码为该按钮绑定了 click 事件处理代码,当"提交"按钮被单击时,通过 CSS 选择器找到 id="saleDetail"表格并判断里面有没有销售明细行元素,如果有就提交表单。代码如下所示。

```
$("#addButton").click(function () {
    let items = $("#saleDetail").find("tr");
    if (items.length > 0) {
        $("#addSaleForm").submit();
```

```
        }
    });
```

11.2 查询销售记录

仔细阅读"查询销售记录"用例描述要求,理解时序图的含义,按要求完成编码工作。

11.2.1 任务需求

查询销售记录用例描述如表 11-2 所示。

表 11-2 查询销售记录用例描述

用例名称	查询销售记录
编号	16
参与者	管理员
简要说明	管理员输入条件查询销售记录
基本事件流	1. 管理员在查询表单中输入开始时间和结束时间,单击"查询"按钮 2. 系统显示符合条件的销售记录
其他事件流	无
异常事件流	返回异常页面,提示错误信息
前置条件	管理员已经登录系统
后置条件	无

查询销售记录功能时序图如图 11-5 所示。

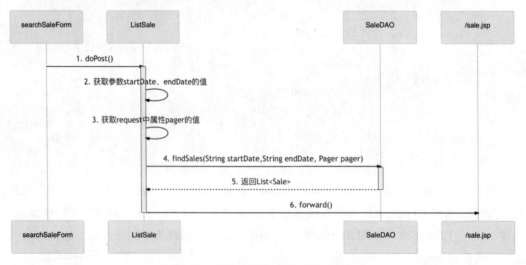

图 11-5 查询销售记录功能时序图

时序图中各参与者说明如下。

searchSaleForm:视图层,放置于 sale.jsp 里面的表单,提交方式为 post,提交地址为

/sale/list,包含控件开始时间 startDate 和结束时间 endDate。

ListSale：控制层，位于 store.controller 包中，是一个 Servlet。其请求地址为/sale/list，处理来自 searchSaleForm 的 post 请求，用于获取查询条件开始时间和结束时间并调用 SaleDAO 的 findSales(String startDate, String endDate, Pager pager)方法进行分页查询，操作完成后返回销售记录查询视图。当单击"上一页"或者"下一页"按钮进行数据查询时，超链接发送的是 get 方式的请求，由于其处理逻辑和表单 post 方式请求完全相同，因此 ListSale 中的 doGet()方法直接调用 doPost()方法即可。

SaleDAO：数据访问模型层。其 findSales(String startDate, String endDate, Pager pager)方法进行分页查询。

/sale.jsp：销售记录管理页面的地址。请求该地址最终会返回销售记录管理视图层。

在销售记录管理主页面的条件输入框中输入查询条件并单击"查询"按钮，分页显示满足条件的查询结果，如图 11-6 所示。

图 11-6　销售记录查询界面

11.2.2　任务实现

【任务 11-6】完成方法 SaleDAO.findSales(String startDate, String endDate, Pager pager)

编辑 store.dao 包下的 SaleDAO 类，并在其中添加方法 findSales(String startDate, String endDate, Pager pager)。

SaleDAO 用来封装对 t_sale 表的操作，其 findSales(String startDate, String endDate, Pager pager)方法进行分页查询，其中 startDate 为查询条件开始时间、endDate 为查询条件结束时间、pager 为分页帮助对象，代码如下所示。

```
public List<Sale> findSales(String startDate,String endDate, Pager pager)
throws Exception {
    List<Sale> list=new ArrayList<>();
    Connection con= null;
    PreparedStatement ps=null;
    ResultSet rs=null;
    try {
```

```java
            Class.forName(DB.JDBC_DRIVER);
            con= DriverManager.getConnection(DB.JDBC_URL,DB.JDBC_USER,DB.JDBC_PASSWORD);
            ps=con.prepareStatement("select count(id) as total from t_sale where trade_time between ? and ?");
            ps.setString(1,startDate);
            ps.setString(2,endDate);
            rs=ps.executeQuery();
            if(rs.next()){
                pager.setTotal(rs.getInt("total"));
            }
            ps=con.prepareStatement("select * from t_sale where trade_time between ? and ? limit ?,?");
            ps.setString(1,startDate);
            ps.setString(2,endDate);
            ps.setInt(3,(pager.getCurrentPage()-1)*pager.getPageSize());
            ps.setInt(4,pager.getPageSize());
            rs=ps.executeQuery();
            while (rs.next()){
                Sale s=new Sale();
                s.setId(rs.getInt("id"));
                s.setTradeTime(rs.getString("trade_time"));
                list.add(s);
            }
    } catch (Exception e) {
        e.printStackTrace();
        throw new Exception("数据库异常:"+e.getMessage());
    }finally {
        if(rs!=null) rs.close();
        if(ps!=null) ps.close();
        if(con!=null) con.close();
    }
    return list;
}
```

【任务 11-7】 完成 ListSale.java

在 store.controller 包下新建 Servlet 类 ListSale,其请求地址为/sale/list。

ListSale 的 doPost()方法用来处理销售记录查询表单的请求,doGet()方法用来处理分页按钮的请求,由于两者的逻辑相同,因此在 doGet()方法中直接调用 doPost()方法即可。ListSale 从请求参数中得到查询条件 startDate 和 endDate,从请求对象的属性中得到分页对象 pager,对 SaleDAO 的 findSales()方法进行调用,得到分页后的查询结果 saleList。查询完毕后通过 request 对象的属性将 pager 和 saleList 两个对象转发到 sale.jsp 页面,代码如下所示。

```java
@WebServlet("/sale/list")
public class ListSale extends HttpServlet {
    protected void doPost(HttpServletRequest request, HttpServletResponse response) throws ServletException, IOException {

        String startDate=request.getParameter("startDate");
        String endDate=request.getParameter("endDate");
        String now=new SimpleDateFormat("yyyy-MM-dd HH:mm:ss").format(new Date());

        if(startDate==null || "".equals(startDate)) startDate=now;
        if(endDate==null || "".equals(endDate)) endDate=now;
```

```
            Pager pager=(Pager) request.getAttribute("pager");
            try {
                List<Sale> saleList=new SaleDAO().findSales(startDate,endDate,pager);
                request.setAttribute("saleList",saleList);
                request.getRequestDispatcher("/sale.jsp").forward(request,response);
            } catch (Exception e) {
                e.printStackTrace();
                throw new RuntimeException(e.getMessage());
            }
        }
        protected void doGet(HttpServletRequest request, HttpServletResponse
response) throws ServletException, IOException {
            doPost(request,response);
        }
    }
```

【任务 11-8】完成视图部分

查询销售记录表单 searchSaleForm 的实现代码位于 sale.jsp 中，以 post 方式提交到 /sale/list，提交参数 startDate 和 endDate 表示要查询的开始时间和结束时间，代码如下所示。

```
<form class="form-inline" action="/sale/list" method="post">
    <label class="mr-1">从</label>
    <input type="date" name="startDate" class="form-control mr-1" />
    <label class="mr-1">到</label>
    <input type="date" name="endDate" class="form-control mr-1" />
    <button type="submit" class="btn btn-success"><i class="bi-search"></i>查询</button>
</form>
```

ListSale 会通过 request 作用域返回两个对象，分别是分页查询结果集 saleList 和分页对象 pager，在 sale.jsp 中通过循环 saleList 生成结果表格，代码如下所示。

```
<table class="table table-hover table-bordered table-striped text-center">
    <thead class="thead-dark">
        <tr>
            <th>序号</th>
            <th>销售日期</th>
            <th class="w-75">明细</th>
        </tr>
    </thead>
    <tbody>
    <c:forEach items="${requestScope.saleList}" var="s" varStatus="vs">
        <tr>
            <td>${vs.count}</td>
            <td>${s.tradeTime}</td>
            <td>
                <a class="btn btn-primary btn-sm" data-toggle="collapse" href="#collapse-${s.id}" load-detail-id="${s.id}">单击查看明细
                </a>
                <div class="collapse" id="collapse-${s.id}">
                    <div class="card card-body">
                    </div>
                </div>
            </td>
```

```
            </tr>
        </c:forEach>
        </tbody>
</table>
```

分页按钮部分代码如下所示,其中超链接请求/sale/list 进行分页查询;传递参数 currentPage 表示希望跳转到第几页,从分页对象 pager 中获取;startDate 和 endDate 为输入的查询条件,从请求参数中获取。

```
<nav>
    <ul class="pagination justify-content-between">
        <li class="page-item">
            <a class="page-link rounded-pill" href="/sale/list?currentPage=${requestScope.pager.previous}&startDate=${param.startDate}&endDate=${param.endDate}"
                tabindex="-1">上一页</a>
        </li>
        <li class="page-item"><a class="page-link rounded-pill">总共${requestScope.pager.pageCount}页,当前第${requestScope.pager.currentPage}页</a>
        </li>
        <li class="page-item">
            <a class="page-link rounded-pill"  href="/sale/list?currentPage=${requestScope.pager.next}&startDate=${param.startDate}&endDate=${param.endDate}">下一页</a>
        </li>
    </ul>
</nav>
```

11.3 查看销售明细

仔细阅读"查看销售明细"用例描述要求,理解时序图的含义,按要求完成编码工作。

11.3.1 任务需求

查看销售明细用例描述如表 11-3 所示。

表 11-3　查看销售明细用例描述

用例名称	查看销售明细
编号	17
参与者	管理员
简要说明	管理员查看销售记录对应的销售明细
基本事件流	1. 管理员在销售记录查询结果列表中单击销售记录对应的查看明细超链接 2. 系统查询销售明细并显示
其他事件流	无
异常事件流	返回异常页面,提示错误信息
前置条件	管理员已经登录系统
后置条件	无

查看销售明细功能时序图如图 11-7 所示。

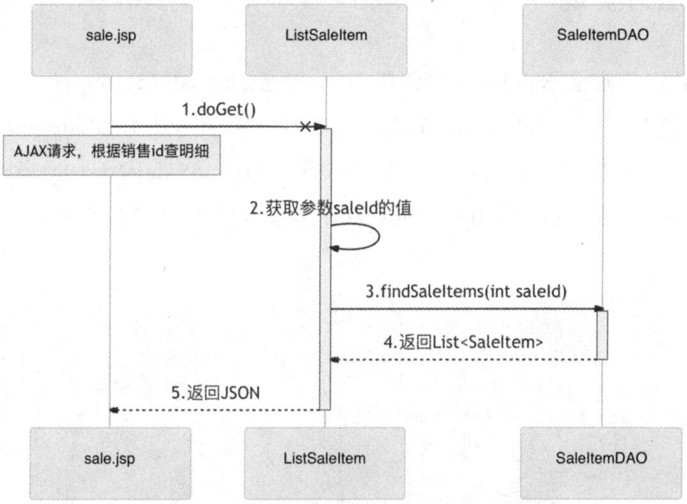

图 11-7　查看销售明细功能时序图

时序图中各参与者说明如下。

sale.jsp：视图层。单击销售记录对应的查看明细超链接，可以异步查询对应销售明细并根据返回的 JSON 数据生成销售明细。

ListSaleItem：控制层，位于 store.controller 包中，是一个 Servlet。其请求地址为/sale/items，处理来自 sale.jsp 的异步 get 请求，调用 SaleItemDAO 的 findSaleItems(int saleId)方法查询指定销售记录的明细，操作完成后以 JSON 格式返回销售明细信息。

SaleItemDAO：数据访问模型层。其 findSaleItems(int saleId)方法根据销售记录主键联合查找 t_sale_item 表和 t_goods 表，将符合条件的销售明细集合返回。

在销售记录查询结果页面单击查看明细超链接，相应销售记录的下方会显示销售明细信息，如图 11-8 所示。

图 11-8　查看销售明细界面

11.3.2 任务实现

【任务 11-9】完成方法 SaleItemDAO.findSaleItems(int saleId)

在 store.dao 包下新建 SaleItemDAO 类，并在其中添加方法 findSaleItems(int saleId)。

SaleItemDAO 用来封装对 t_sale_item 表的操作，其 findSaleItems(int saleId)方法根据销售记录主键联合查询 t_sale_item 表和 t_goods 表，将查询结果封装成 SaleItem 集合返回，代码如下所示。

```java
public class SaleItemDAO {
    public List<SaleItem> findSaleItems(int saleId) throws Exception{
        List<SaleItem> list=new ArrayList<>();
        Connection con= null;
        PreparedStatement ps=null;
        ResultSet rs=null;
        try {
            Class.forName(DB.JDBC_DRIVER);
            con= DriverManager.getConnection(DB.JDBC_URL,DB.JDBC_USER,DB.JDBC_PASSWORD);
            ps=con.prepareStatement("select si.count,g.* from t_sale_item as si inner join t_goods as g on si.goods_id=g.id where si.sale_id=? ");
            ps.setInt(1,saleId);
            rs=ps.executeQuery();
            while (rs.next()){
                SaleItem si=new SaleItem();
                si.setCount(rs.getInt("count"));
                Goods g=new Goods();
                g.setId(rs.getInt("id"));
                g.setName(rs.getString("name"));
                g.setSpecs(rs.getString("specs"));
                g.setSn(rs.getString("sn"));
                g.setPrice(rs.getDouble("price"));
                g.setStock(rs.getInt("stock"));
                g.setCategoryId(rs.getInt("category_id"));
                si.setGoods(g);
                list.add(si);
            }
        } catch (Exception e) {
            e.printStackTrace();
            throw new Exception("数据库异常:"+e.getMessage());
        }finally {
            if(rs!=null) rs.close();
            if(ps!=null) ps.close();
            if(con!=null) con.close();
        }
        return list;
    }
}
```

【任务 11-10】完成 ListSaleItem.java

在 store.controller 包下新建 Servlet 类 ListSaleItem，其请求地址为/sale/items。

ListSaleItem 的 doGet()方法接收前端的 AJAX 请求，然后通过 SaleItemDAO 查询指定销售记录主键的销售明细集合，最后以 JSON 格式返回数据。代码如下所示。

```java
@WebServlet("/sale/items")
public class ListSaleItem extends HttpServlet {
    protected void doGet(HttpServletRequest request, HttpServletResponse response) throws ServletException, IOException {
        int saleId=Integer.parseInt(request.getParameter("id"));
        try {
            List<SaleItem> saleItemList=new SaleItemDAO().findSaleItems(saleId);
            JSONObject jsonObject=new JSONObject();
            jsonObject.put("success",true);
            jsonObject.put("data",saleItemList);
            response.setContentType("application/x-json");
            response.getWriter().write(jsonObject.toJSONString());
            response.getWriter().flush();
            response.getWriter().close();
        } catch (Exception e) {
            e.printStackTrace();
            JSONObject jsonObject=new JSONObject();
            jsonObject.put("success",false);
            jsonObject.put("msg",e.getMessage());
            response.setContentType("application/x-json");
            response.getWriter().write(jsonObject.toJSONString());
            response.getWriter().flush();
            response.getWriter().close();
        }
    }
}
```

【任务 11-11】完成视图部分

在生成销售记录列表的时候，每一行记录都有一个查看明细超链接，代码如下所示。

```html
<a class="btn btn-primary btn-sm" data-toggle="collapse" href="#collapse-${s.id}" load-detail-id="${s.id}">单击查看明细
</a>
```

代码中 data-toggle="collapse"用来指定 Bootstrap 的触发行为是折叠，href="#collapse-${s.id}"用来指定当前超链接对应的折叠层 id，对应下面代码中的 id="collapse-${s.id}"，load-detail-id="${s.id}"用来让 jQuery 取到对应的销售 id。

```html
<div class="collapse" id="collapse-${s.id}">
    <div class="card card-body">
    </div>
</div>
```

通过 jQuery 为含有 load-detail-id 属性的元素绑定单击事件，在事件处理中完成对/sale/items 的 AJAX 请求，并根据返回的 JSON 数据生成销售明细，最后以折叠的方式显示出来。代码如下所示。

```javascript
$("[load-detail-id]").click(function () {
    let sale_id = $(this).attr("load-detail-id");
    $.getJSON("/sale/items", {id: sale_id}, function (result) {
        let collapse = $("#collapse-" + sale_id);
        collapse.empty();
```

```javascript
        let html = "<table class='table table-condensed  table-sm'>" +
            "<tr><td>商品</td><td>单价</td><td>数量</td></tr>"
        let saleItems = result.data;
        saleItems.forEach(item => {
            html += '<tr>
                            <td>\${item.goods.name}</td>
                            <td>\${item.goods.price}</td>
                            <td>\${item.count}</td>
                    </tr>'
        });
        html += "</table>"
        collapse.html(html);
    })
});
```

本章习题

1. 结合项目代码，实现添加销售记录功能。
2. 结合项目代码，实现查询销售记录功能。
3. 结合项目代码，实现查看销售明细功能。